RECENT ADVANCES IN DEVELOPMENTAL BIOLOGY

Recent Advances in Developmental Biology

Vol. 9
Animal Regeneration

A.P. DIWAN
N.K. DHAKAD

ANMOL PUBLICATIONS PVT LTD
New Delhi-110 002

ANMOL PUBLICATIONS PVT LTD
4374/4B, Ansari Road
Daryaganj, New Delhi-110 002

Recent Advances in Developmental Biology

First Edition 1995
ISBN 81-7488-039-9

PRINTED IN INDIA

Published by J.L. Kumar for Anmol Publications Pvt. Ltd. and
Printed at Mehra Offset Press, Delhi.

Preface

The fundamentals of Developmental Biology are presented within the framework of scientific discovery. Researches in Embryology have made almost incredible strides in the past few decades. Consequently, the existing concepts of the developmental biology have expanded. There has been a revolution indeed in this direction.

The text integrates the descriptive, experimental and biochemical approches into a conceptual framework for the analysis of development. All important points are illustrated diagrammatically. The title is not intended to be comprehensive, nor could it be at this length, but concentrates as putting across the basic principles of the subject as briefly and lucidly as possible. It does this with the aid of careful selected examples, some recent and other classic of the field and with numerous illustrations. The aim is to enthuse the reader with this active and exciting area of research and to lay a solid foundation on which further study of its various facts may be based.

Authors

Contents

1

Regeneration

Regeneration is basically the recovery of lost part. In lower animals including invertebrates it is thought of as a form of asexual reproduction. This is particularly true in plants; the formation of an entire plant from one organ, such as a leaf, is an example of regeneration and just as certainly represents asexual reproduction. Cap formation in *Acetabularia* could be treated under the heading of regeneration. This chapter deals with animal systems of regeneration. Rarely do animals reproduce by asexual means, exceptions being the protozoa and budding found in coelenterates, polychaets, and ascidians. However, many animals are capable of replacing lost parts and, in some cases, are capable of producing a complete pair of animals from a single individual. In planarian regeneration, it represents a system capable of reproducing a complete pair of planarians from one parent individual. Where is the regeneration of amphibian limbs it is restricted to the appendages (forelimbs and hindlimbs, tail, jaws) of all larvae and of adult urodeles. An indication of the kinds of results obtained with mammalian limbs is given following the discussion of the urodele limb.

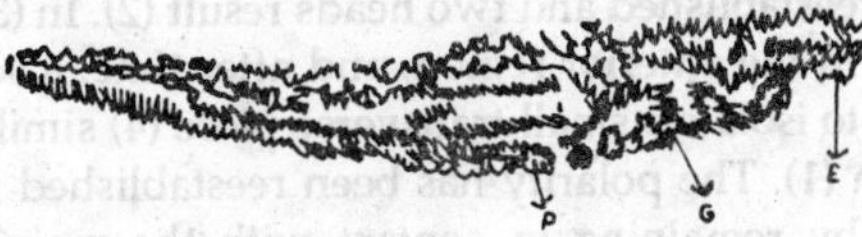

Fig. 1.1. Photograph of fixed and strained adult planarian. E, eye; G, gut; P, pharynx.

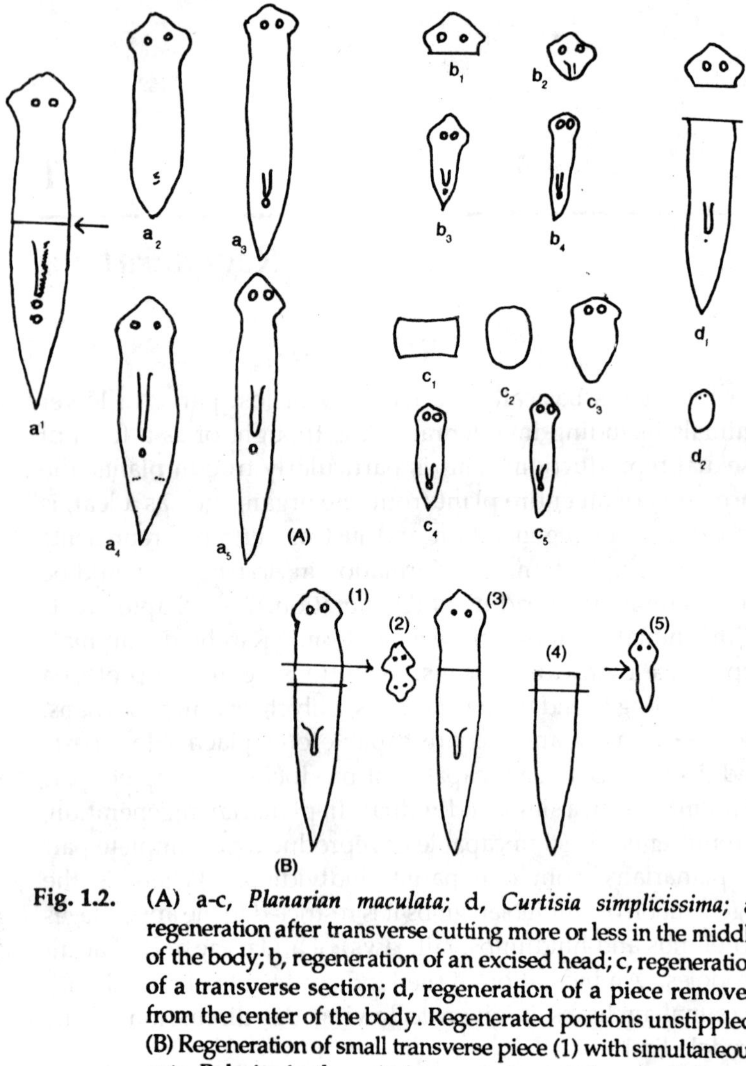

Fig. 1.2. (A) a-c, *Planarian maculata;* d, *Curtisia simplicissima;* a, regeneration after transverse cutting more or less in the middle of the body; b, regeneration of an excised head; c, regeneration of a transverse section; d, regeneration of a piece removed from the center of the body. Regenerated portions unstippled. (B) Regeneration of small transverse piece (1) with simultaneous cuts. Polarity in the anterior-posterior axis has not had time to be reestablished and two heads result (2). In (3), one cut is made in the anterior region, and after time, a second cut is made to isolate a small transverse piece (4) similar in size to that in (1). The polarity has been reestablished in the small piece by remaining in contact with the main body after decapitation, and the small transverse piece in this case develops into a normal worm.

REGENERATION IN PLANARIA

The Animals

Adult planaria are easily maintained in the laboratory and can be collected or purchased through supply houses. The size of the adults depends on the species, but generally runs about • inch in length for those used in the laboratory. Not all representatives of the platyhelminthes have regenerative power. The triclads (including planaria) have a most striking capacity for regeneration under a wide variety of experimental conditions.

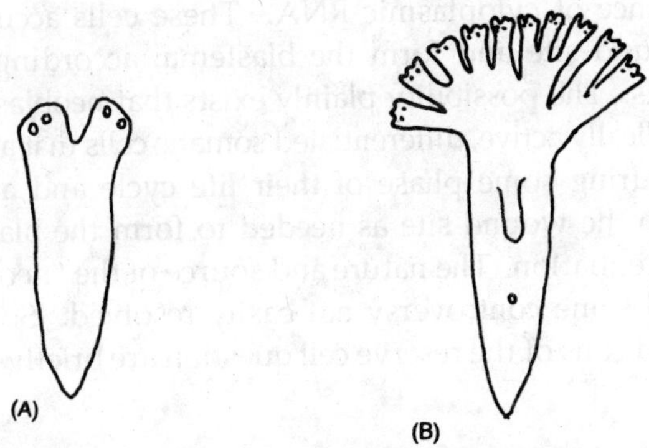

Fig. 1.3. Multiple head regeneration: (A) twin heads after decapitation and a medial incision in Curtisia simplicissima; (B) numerous heads formed after decapitation and numerous incisions in the anterior part of the body in *Dendrocoelum lacteum*.

When a planarian is cut in half transversely, regeneration is initiated at both wound surfaces. The anterior end grows a new tail and the posterior end, a new head. The missing parts develop from regeneration blastemas which are collections of undifferentiated cells that accumulate at the wound surfaces. In lower forms (coelenterates), regeneration involves the reorganization of existing differentiated cells and tissues into new parts — a process called morphoallaxis — which is in contrast to regeneration via a blastema where mitosis produces new cells.

BLASTEMAL CELLS

The Neoblast Question

There are theoretically two ways to produce undifferentiated cells: either by dedifferentiation and recommitment of existing tissues, or by maintaining a supply of reserve cells which remain uncommitted throughout embryogenesis and are thus available for use in later life. The latter alternative may be followed by planarians. Cells called neoblasts are distributed throughout the body. They are distinguished mainly by an abundance of cytoplasmic RNA. These cells accumulate at the wound site and form the blastema, according to many accounts. The possibility plainly exists that neoblasts may be metabolically active, differentiated somatic cells that accumulate RNA during some phase of their life cycle and are able to move to the wound site as needed to form the blastema, by dedifferentiation. The nature and source of the "neoblast" has aroused some controversy not easily resolved. Some of the pros and cons of the reserve cell question are briefly discussed below.

The Source of Blastemal Cells

Normally, the cells of the blastema are of local origin in the vicinity of the wound. Neoblasts are supposedly more susceptible to X-radiation than are other somatic cells. Irradiation of a band of tissue at right angles to the long axis of the worm destroys the neoblasts of that region. When a cut is made through the irradiated zone, regeneration is delayed beyond the non-irradiated control times. Histological studies indicate that this longer time period is required for the migration of neoblasts from unirradiated portions of the body to the site of amputation. The loss of neoblasts from irradiated portions and the recolonization, presumably from other healthy body areas, has been documented. Whole body irradiation of the worm leads to loss of neoblasts from the entire body as well as to complete loss of regenerative powers. In planarians that have been totally irradiated to inhibit regeneration, this capacity

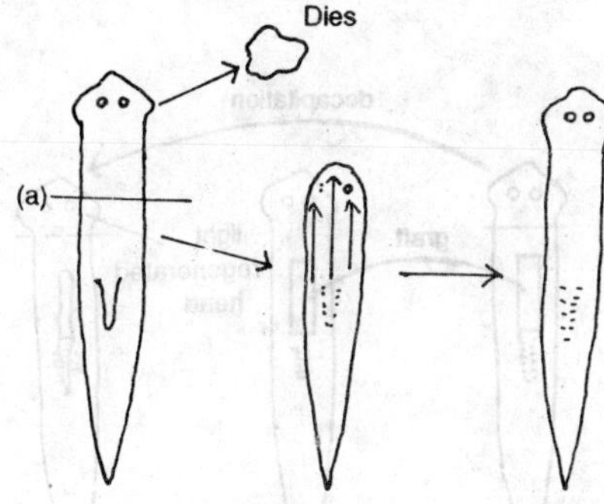

Fig. 1.4. Effects of irradiation on regeneration. Shaded portion, irradiated. Clear portion, shielded from x-rays. Cut made at (a) following irradiation. Irradiated head dies. The anterior end regenerates, after delay, by migration of cells from the shielded portion to form a blastema at the anterior end.

may be restored by a graft of healthy tissue from a non-irradiated worm. Cutting the treated planarian above the region of the graft results in the migration of healthy graft cells into the irradiated portion of the host, the formation of a blastema, and regeneration of the missing part. If the graft is marked by more lightly pigmented cells, the regenerated parts will reflect the pigmentation of the graft. The neoblasts of the graft are claimed to be the essential cells contributed by the graft. The experiment allows one to rule out reactivation of irradiated host tissue by the healthy graft tissue as a regeneration mechanism, though it is possible that a few of the host cells may participate in a limited way in the formation of the blastema. These studies would seem to support the existence of a population of undifferentiated reserve neoblasts.

An alternative explanation exists, X-ray treatment may also act to prevent the dedifferentiation of somatic cells that would also lead to failure of blastema formation. The existence of a completely uncommitted population of reserve cells remains unproved. The problem is complicated by the fact that all blastemal cells look alike, and identification of those cells

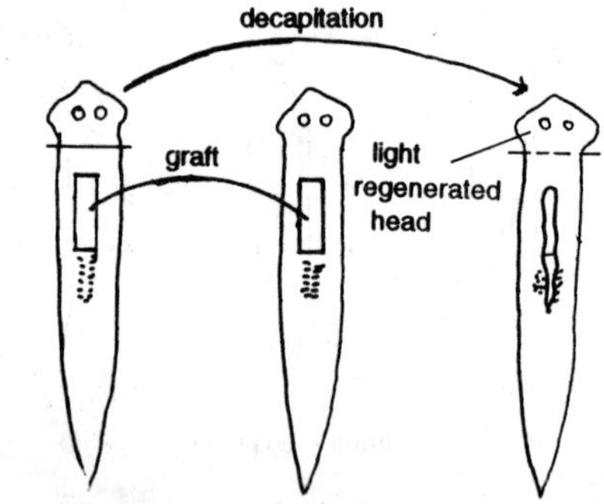

Fig. 1.5. Regeneration after total body irradiation by implantation of tissue graft from non-irradiated donor of another species (clear block). After the graft is in place, the host is decapitated. Regeneration follows utilizing cells from the non-irradiated donor. The regenerated head is of the donor pigmentation.

destined to become blastemal cells is certain only after they have entered the blastema. With their prior identity lost, it is at best difficult to be certain of the tissue origins of blastemal cells. Some planarian species possess neoblasts, but are incapable of regeneration. Possibly, the movement or proliferation of the undifferentiated cells in these planarians is restricted, thus precluding accumulation of these cells at the wound. It is also possible that such cells are not involved in regeneration. In any event, a blastema of undifferentiated cells is formed, regardless of the origin of the cells, either as undifferentiated neoblasts or as dedifferentiated cells. Muscle dedifferentiation has been observed in planarian regeneration, and it is probable that both neoblasts and dedifferentiated cells contribute to the blastema and the resulting regenerate. For convenience, all cells forming the blastema will be referred to as *"neoblasts"* regardless of possible origins.

Differentiation of the Blastema

After transection of the worm, blastema formation takes

only a few days, and a week or two will suffice for the appearance of a new head or tail structure. Areas of the blastema concerned with head regeneration on the posterior portion have distinct fats. Those cells in the middle become nerve cells and give rise to the brain. Other cells in other locations give rise to eye spots, muscle, intestine, excretory, or reproductive tissues. Some of the blastemal cells may possibly remain as neoblasts, in the undifferentiated state.

Pluripotency of "Neoblasts"

Since "neoblasts" from one region of the body can apparently be made to populate another region, giving rise to normal regeneration, it has been concluded that the neoblasts do not possess regional specificity. Rather, the neoblasts are pluripotent and their fate is impressed on them by the tissues surrounding the blastema. A striking example of local determination of neoblast fate is found in the simplest experiment: cutting a planarian into an anterior and posterior half. The neoblasts appearing at the cut edge of the wound were originally located at the same body level. However, those of the anterior half will give rise to a tail and those of the posterior half will give rise to a head, two clearly distinct structures. This observation leads to a consideration of polarity in planaria. Polarity in this instance reflects the ability of the head end to correctly produce a tail and of the tail section to correctly produce a head.

POLARITY IN PLANARIAN REGENERATION

The Role of Gradients

As in many developing systems (e.g., *Acetabularia* an anteroposterior gradient has been postulated to account for head and tail determination. The nature of the gradient has yet to be conclusively established, but evidence again points to involvement of some species of RNA being distributed in gradient fashion along the length of the worm. Following the gradient hypothesis, some interesting experiments have been

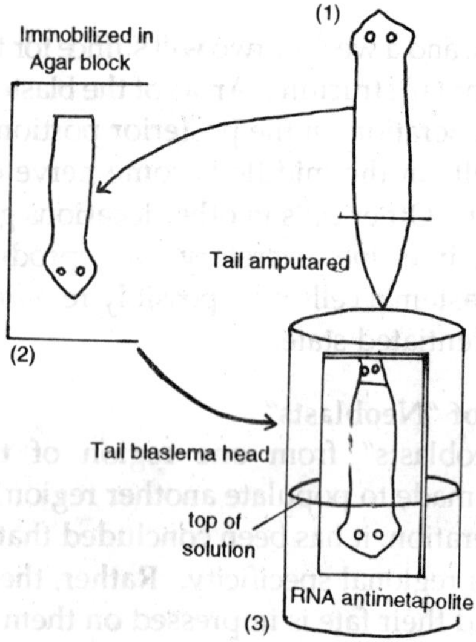

Fig. 1.6. Reversal of polarity by antimetabolites interfering with RNA synthesis. (1) The tail is amputated, and discarded. (2) the tailless worm is embedded in an agar block to immobilize it, and then immersed (3), head end down, into the antimetabolite. The tail section is not immersed in the solution. The result is a head regenerating on the tail end. This experiment provides indirect evidence for involvement of RNA in establishment of the postulated anterior-posterior polarity gradient.

carried out. Very short pieces of the worm, derived from two simultaneous horizontal slices, do not have sufficient gradient differential across their length to distinguish head and tail gradient concentrations. These pieces develop heads at both ends. If the second cut producing the small slice is delayed, polarity will have become established and the second (posterior) head end will be supressed in favour of a tail structure. These results are consistent with a gradient hypothesis. They do not established the hypothesis as fact. Actually, the statement of the hypothesis is little more than description of the events observed during the course of the experiment, with the supposition of a gradient added as an "explanation" of the happenings.

An RNA gradient ? Some evidence is available in support of the theory stating that the gradient is one involving RNA concentration or rates of RNA synthesis. Polarity may be reversed by immersing the head end in antimetabolites, which suppress RNA synthesis as well as the synthesis of other molecules (e.g., DNA), thus interfering with the postulated gradient. The amputation of the tail end of a worm treated in this manner leads to the regeneration of a head at the posterior end. Since the original head recovers following the treatment, a bipolar two-headed monster is produced.

DETERMINATION OF ISOLATED BLASTEMAS

Separation of head or tail blastemas from the parent body demonstrates that from the time of formation, the blastema has been determined to form a particular structure. The head and tail blastemas form, respectively, only heads or tails of small size following isolation. Two complementary blastemas can be grafted together to produce a complete worm. The qualitative determination of the blastema is seemingly irreversible in the face of many different kinds of experimental grafting and surgical procedures.

TIME COURSE OF BLASTEMA DETERMINATION

Stepwise Determination

The design of an experiment illustrating stepwise determination is outlined. In the head blastema, the brain and eyes form from the middle third of the neoblasts on the midline of the body. The experiment shown is designed to remove this midportion at progressively later times during head regeneration. Normally, as regeneration of the head progresses, the fate of the lateral cells is to form tissues other than brain and eyes. If the lateral cells are isolated from the central cells of the blastema early in regeneration, the lateral cells will form brain and eyes after a slight delay. As regeneration proceeds, the lateral thirds of the blastema may still form brain and eyes when isolated, but progressively longer times are required for them to do so. In other words, the longer the original blastema remains intact, the more limited the lateral portions become with respect to their ability to form both brain and eyes, though they eventually succeed in doing so given time for reorganization. The results suggest the presence of an inhibitor of brain formation that spreads, slowly from the early determined central brain cells to the presumed lateral eye cells as regeneration proceeds. The neoblasts are thus gradually (but reversibly) restricted or determined in their potential by a negative influence emanating from more central cells. This type of inhibitory influence can be demonstrated in other ways. For example, a head grafted in the vicinity of the existing head will be resorbed, while a head grafted to the tail region survives. The grafted head will also survive in the head region if the host head is removed after the graft has been made. Regeneration of the original host head is then inhibited by the presence of the grafted head.

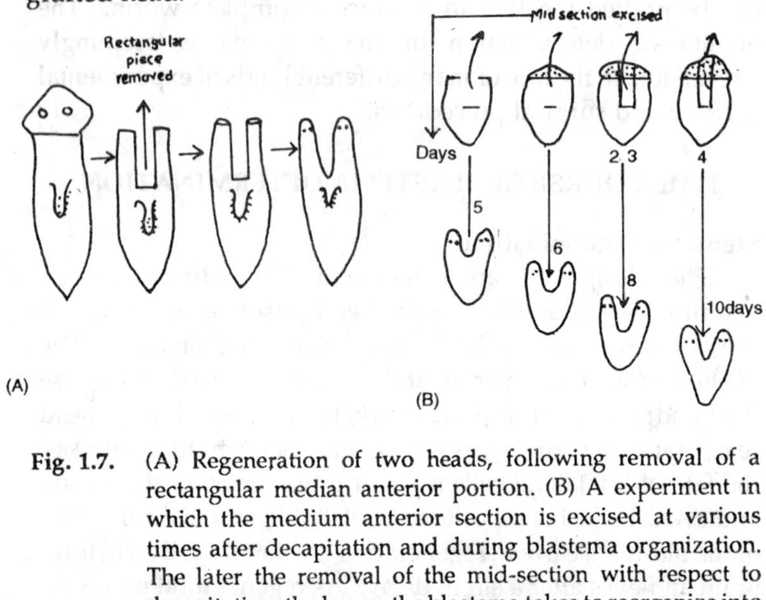

Fig. 1.7. (A) Regeneration of two heads, following removal of a rectangular median anterior portion. (B) A experiment in which the medium anterior section is excised at various times after decapitation and during blastema organization. The later the removal of the mid-section with respect to decapitation, the longer the blastema takes to reorganize into two new head blastemas. This is indicative of a stepwise determination in the regeneration blastema.

Possible Induction

In other instances, a positive or inductive influence seems to be operating. In general, eyes are dependent on the presence of a brain for their formation. If the eyes are removed, they will regenerate. If the eyes and brain are removed simultaneously, the eyes will not reappear in the absence of the brain.

The "inductive" effect of the brain extends to other species of flat-worms, which are able to support differentiation of eyes in brainless worms after a heterograft procedure placing the foreign brain in the brainless planarian. Brain tissue extracts will also induce eyes, though tail extracts do not. The brain seems to be the initiator of determination, which may, subsequently, influence the development, in a stepwise fashion, of the other parts of the blastema. This stepwise determination also includes a negative influence on, or restriction of, developmental potential in the various regions of the blastema.

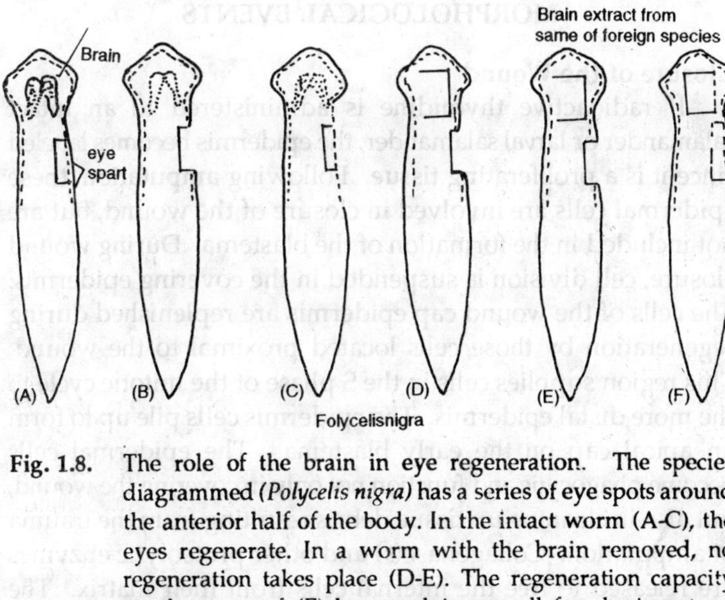

Fig. 1.8. The role of the brain in eye regeneration. The species diagrammed *(Polycelis nigra)* has a series of eye spots around the anterior half of the body. In the intact worm (A-C), the eyes regenerate. In a worm with the brain removed, no regeneration takes place (D-E). The regeneration capacity can be restored (E) by supplying a cell free brain extract (even from a different species).

Regeneration of the Larval or Adult Salamander Limb

The development phenomena in regeneration of the amphibian limb are similar to those encountered in the development of the embryonic limb, once the blastema has been established. The origin of the mesenchymal cells in the two cases is, of course, quite different. The embryonic limb mesenchyme arises from cells of the somatic mesoderm. The cells of the regeneration blastema are derived from the original limb tissues by a process of dedifferentiation. The involvement of reserve cells in amphibian regeneration appears to be minimal. The existence of reserve cells as distinct cytological entities has not been seriously proposed in recent years.

Regeneration in amphibians is limited to the larval stages of anuran and urodele species and to adult urodeles. *Xenopus* represents the exception among the anurans in that adults of this genus have the capacity for limited regeneration of appendages.

MORPHOLOGICAL EVENTS

Closure of the Wound

If radioactive thymidine is administered to an intact salamander or larval salamander, the epidermis becomes labeled since it is a proliferating tissue. Following amputation, these epidermal cells are involved in closure of the wound, but are not included in the formation of the blastema. During wound closure, cell division is suspended in the covering epidermis. The cells of the wound cap epidermis are replenished during regeneration by those cells located proximal to the wound. This region supplies cells in the S phase of the mitotic cycle to the more distal epidermis. The epidermis cells pile up to form an apical cap on the early blastema. The epidermal cells become phagocytic and function not only in covering the wound, but also in cleaning up tissue debris resulting from the trauma of amputation. Collagenase(s) and other proteolytic enzymes are released to free the internal cells from their matrix. The epidermal cells may be, in part, responsible for production of these enzymes. With the closure of the wound, dedifferentiation of internal tissues and blastema formation begin.

Histogenesis of the Blastema

The blastema arises from the mesodermal elements of the stump. All tissues, including cartilage, muscle, and dermal elements, are able to participate. This local origin of blastemal cells was again established by use of X-radiation. Local irradiation inhibits regeneration and blastema formation. Whole body irradiation, with the amputation site shielded, does not.

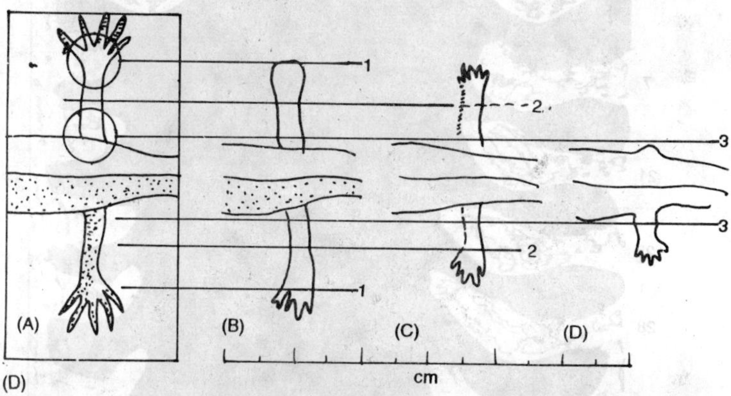

Fig. 1.9. (D) Local irradiation of parts of the limb in a newt, and results of amputation of different levels in the irradiated limb (right) and in control limb (left). Drawing at left (A) shows shielding plate covering the whole animal except for two circular openings over right forelimb through which the x-rays could pass. The limb fails to regenerate when amputated inside irradiated areas (lines 1 and 3). Control (left) side regemerates at all levels.

After amputation, cells of the stump tissues lose their histological identity and become wandering mesenchyme cells. Syncytial muscle breaks down into individual mononucleate cells, without striations. Cartilage cells lose their matrix of collagen. Redifferentiation of the blastema takes place in a proximodistal direction. Blastema formation and regeneration take place in similar fashion in both larval and adult urodeles, though the larvae achieve total replacement much more rapidly. The more proximal the amputation, the more rapid the regeneration of either larval or adult limbs. This relationship was noted in the eighteenth century by an Italian biologist named Spallanzani. It is still not understood.

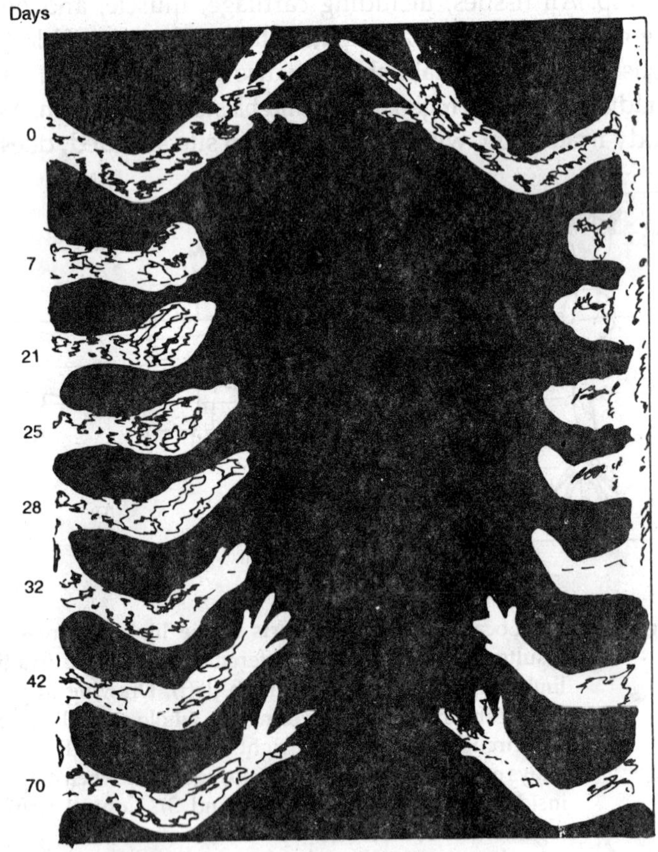

Fig. 1.10. Successive stages in the regeneration of opposite limbs in a newt following amputation through lower (left) and upper (right) arms. At the top are the original limbs. Reading down, photographs were taken at 7, 21, 25, 28, 32, 42 and 70 days after amputation. From more proximal stumps (right) the regenerate elongates faster but differentiates more slowly than that growing from more distal level of amputation.

After the mesenchymal cells have formed the blastema, the developing appendage assumes a conical shape. The inner cells remain undifferentiated, while the epidermis covering the distal portion becomes thickened. The epidermis assumes a shape similar to the apical cap of the embryonic limb bud. The ensuing cytodifferentiation closely resembles that observed in limb development. The mesenchymal cells destined to form cartilage align themselves along the midline of the regenerating structure, beginning in the proximal portion of the blastema. The regenerate becomes paddle-shaped. The apical epidermis then thins and associates closely with the developing phalanges (finger-buds). During the finger-bud stage of regeneration, the cartilages of the limb are defined. The cartilagenous matrix is secreted as the precartilage cells pass through a previously described cycle of structural and metabolic changes, including hypertrophy of cell organelles, differentiation of tissues, and suspension of growth and mitosis. Muscle develops in a proximodistal direction adjacent to the cartilage.

The Essential Role of the Epidermis

In a manner similar to the influence of the ectodermal cap of the frog limb bud in limb differentiation, the apical cap of the regenerating limb appears to be essential for blastema differentiation. Daily removed of the apical epidermal cap inhibits the production of a blastema. If the epidermal cap is removed and replaced in a more lateral position, the blastema will develop under the epidermis is an abnormal orientation, skewed from the original long axis of the amputated limb. The normal epidermis of the blastema can be replaced with epidermis from other parts of the body. The limbs resulting from such epidermal grafts are normal. Indicating that the epidermis does not specify the kind of regeneration. The information regarding forelimbs versus hindlimbs or tail is a function of the mesenchymal blastema. However, an epidermal covering is necessary for blastema formation and continued differentiation.

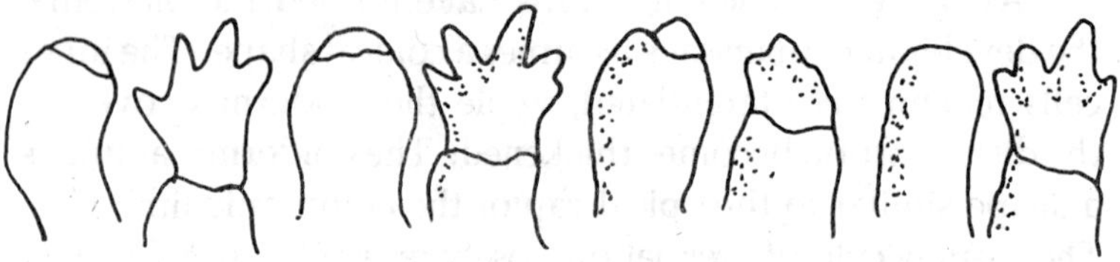

Fig. 1.11. Arrest of regeneration in limbs with transected nerves (left limb in each pair). Limbs with intact nerves regenerate normally.

The Role of Nerves in Regeneration

Prevention of nerve outgrowth into the regenerating stump effectively prevents regeneration. A central connection of the peripheral nerves is not required, sensory nerves severed on the spinal cord side of the dorsal root ganglion are effective in supporting limb regeneration. A minimum, or threshold, number of nerve fibers supplying the limb is necessary to support regeneration. Limbs containing fewer than $^1/_3$ to $^1/_2$ of their normal number of nerve fibers fail to regenerate following amputation. During embryonic development the limbs differentiate in advance of innervation; a nerve supply is not required in limb-bud differentiation. Similarly, limbs which have never become innervated, in larvae surgically deprived of the neural tube, will regenerate in the absence of a nerve supply. However, once the aneurogenic limbs have become innervated through implanted nerves, a nerve supply is required for regeneration. Experiments point to contact with the epidermis of the early cone-shaped blastema as the point of influence of neuronal factors. A lack of infórmation exists regarding the nature of possible trophic substances produced by nerves during regeneration. It remains that acetylcholine has been shown not to be the substance involved in neuronal control of regeneration.

The Role of Endocrines in Regeneration

Regeneration can be inhibited by prior removal of the pituitary gland. If replacement therapy is instituted, it is found that either ACTH or cortisone is effective. The pituitary appears to be necessary only in so far as it is necessary for the activity of the adrenal cortex. Administration of drugs that inhibit cortisone secretion also inhibits regeneration. This block may be relieved by cortisone, but not by pituitary hormone (ACTH). The role of cortisone is apparently limited to early phases of wound healing. The formation of a blastema does not appear to depend on cortisone. Cortisone does interfere with dermal scarring of the wound, which occurs by deposition of collagenous connective tissue fibers. Such scarring may prevent normal interactions between epidermis and mesoderm that are necessary for regeneration to take place.

As a rule, larval regeneration is less dependent on adrenal secretions than is regeneration in adult salamanders.

REGENERATION FIELDS AND POLARITY

Blastema and Limb Bud

Comparisons between the embryonic limb bud and the regeneration blastema reveal that the two systems have much in common, but they are by no means-identical. The limb bud has the potential to develop the entire limb. The blastema is limited to those structures of the limb or tail distal to the amputation. The blastema conforms to the morphological pattern of the parent stump as regeneration proceeds, while the limb bud develops as an entire limb field. The type of limb to be developed (forelimb or hindlimb) is a function of the mesodermal portion of the limb bud or of the mesenchyme of the blastema. A tail blastema transplanted to the area of an amputated limb will develop a combination tail-limb. If the implantation site is irradiated to destroy the cells normally forming the limb blastema, only a tail will be formed from this transplant.

Instructions from the Stump

Experiments involving rearrangements of limb skeletal parts deal with the effect of stump tissues on regeneration. If the humerus of the upper arm is removed disarticulated) and the limb subsequently amputated through the upper arm, the

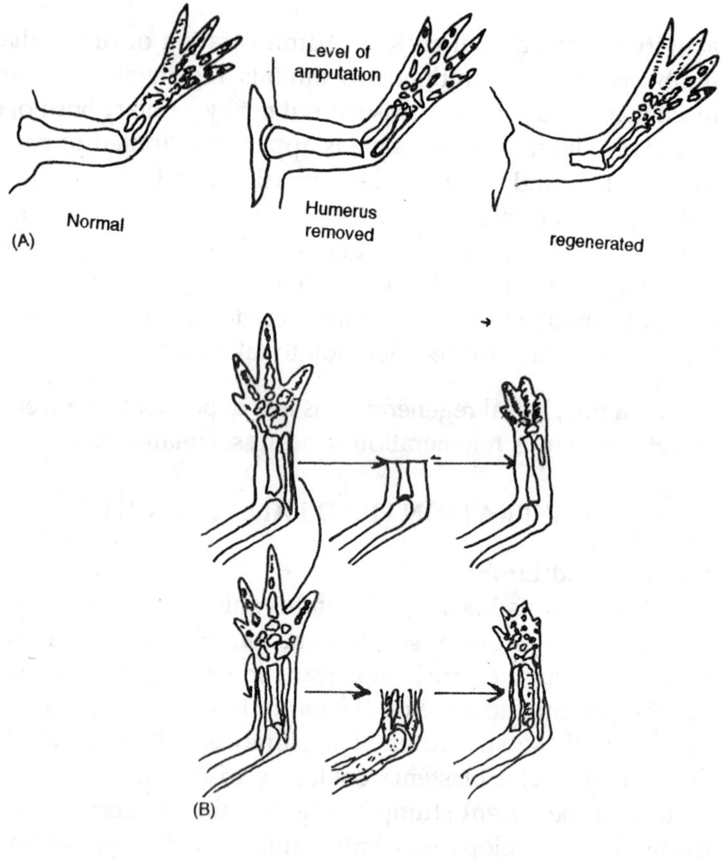

Fig. 1.12. (A) Regeneration of a forelimb in a newt from which the humerus has been previously removed (diagrammatic). (B) Addition of an extra ulna to a lower arm, followed by amputation, brings about the production of a corresponding extra skeletal element in the regenerate.

distal portion of the humerus is produced in the regenerate. The proximal portion is never replaced (regeneration always proceeds in a direction distal to the amputation). Similarly, removal of the ulna from the forearm, followed by amputation through the forearm, results in the formation of the ulna distal to the amputation. In a similar fashion, an extra ulna may be inserted in the forearm prior to amputation. The regenerate will faithfully reproduce the extra element during regeneration. The latter instance would favour the view that the tissues of the stump are responsible for directing the pattern of regeneration. This conclusion must be modified in the case of the regeneration of the missing ulna or humerus, since no corresponding skeletal element is present to direct this regeneration. The limb is capable of replacing skeletal deficiencies during regeneration, yet has no mechanism for eliminating supernumary elements. In any event, the prior organization of the limb stump is clearly an important directing factor. The presence of the skeletal elements as such is not the only factor in organization of the skeletal elements during replacement. The surrounding tissues must contribute spatial instructions to the blastema, as well as cells (other than those of the skeleton) for the formation of the distal portion of the humerus found in the regenerate.

Self-differentiation of the Blastema

It is possible to assess the role of stump tissues by growing the blastema in isolation from the stump. Blastemas in the cone stage of development can continue their development in the absence of the stump. Cartilage and muscle differentiate within these blastemas, and skeletal elements of the arm and hand are produced. The morphology is distorted, due to the lack of surrounding body tissues and to the lack of a blood supply. The blastema at this stage has received the influence from nerves, as well as perhaps some instructions from the stump. Explanation of the apical blastema to the flank or tail fine region gives rise to distal limb structures, which are vascularized and well formed. Since the symmetry is normal in these ectopic limbs, it is presumed that some early instruction form the stump has been imparted and utilized in organization of the explanted, stumples blastema.

Regeneration from Anterior Halves

If the posterior portion of the forelimb of a salamander is cut away from wrist to elbow, subsequent amputation produces a stump of the anterior half only. From such a bisected half-limb, only a half-limb regenerates. If only the internal tissues are removed (muscle, bone) leaving the epidermis and dermis intact a complete limb will regenerate from the amputated stump. This points to the dermis as a possible source of instructions to the blastema.

Reversal of Polarity

The usual proximodistal direction of regeneration can be reversed in the sense of the direction of regeneration with respect to the limb itself. Here, the wrist portion of the forearm of a salamander is grafted to the flank after amputation of the hand. Following healing, the forelimb is severed through the upper arm. Regeneration occurs from both cut surfaces, resulting in distal structures in both cases. In one instance, the direction of regeneration has been reversed with respect to the limb axes, but remains proximodistal with respect to the body of the animal.

The Question of Differentiation and Pluripotency of Blastemal Cells

All cells types found in the limb contribute to the formation of the blastema, with the possible exception of epidermal and blood cells. The question at hand concerns the potential for redifferentiation of the variously derived cells of the blastema. For example, are cartilage cells capable of joining the blastema and reemerging as muscle cells, or is their fate restricted to cartilage ?

The critical experiment consists of grafting marked (triploid) tissues into salamander limbs prior to amputation. The triploid cells are recognizable since they carry three nucleoli in the interphase nucleus instead of two. The triploid grafts used were pure cartilage, *"pure"* muscle, cartilage including the mitotically active surrounding membrane (the perichondrium), and epidermis. Host limbs were irradiated prior to grafting and amputation to discourage participation of diploid host cells in the regenerate.

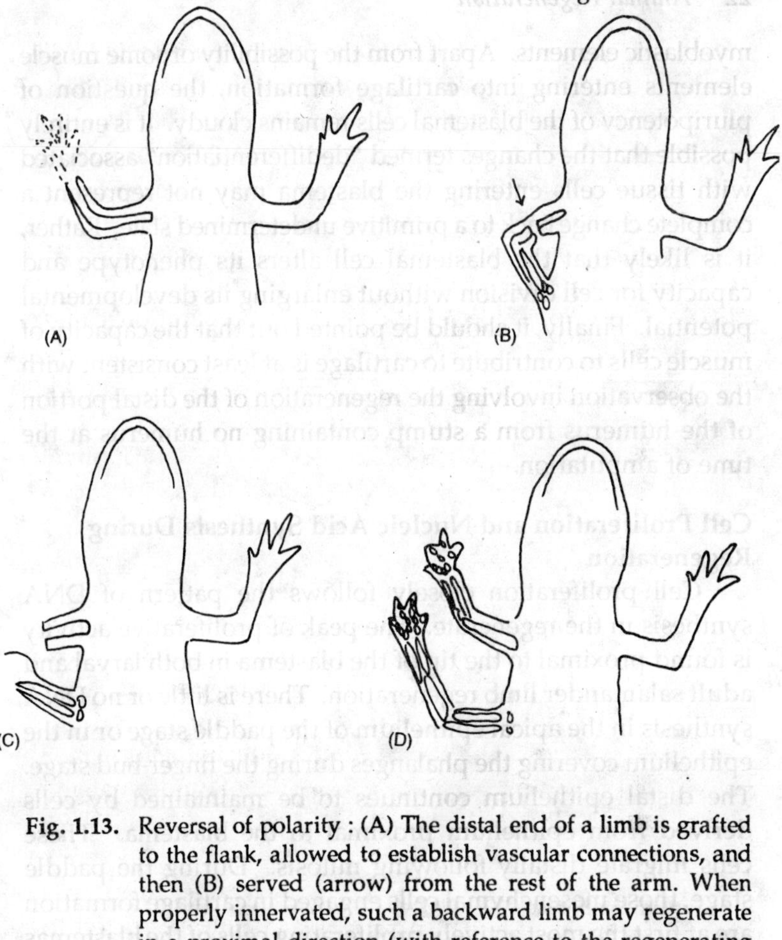

Fig. 1.13. Reversal of polarity : (A) The distal end of a limb is grafted to the flank, allowed to establish vascular connections, and then (B) served (arrow) from the rest of the arm. When properly innervated, such a backward limb may regenerate in a proximal direction (with reference to the regenerating limb) but will always form distal structures in doing so. (C-D)

When grafts consists of pure cartilage or cartilage plus perichondrium, the triploid cells appeared in cartilage, perichondrium, and connective tissue, but not in muscle tissue. When "pure" muscle is used as the graft, the marker cells appear in all the tissue types as well as in muscle cells. Epidermal graft cells do not contribute to any tissues in the regenerate. When considering muscle, it is important to recall that muscle consists of connective tissue and blood tissues as well as myoblastic elements. Apart from the possibility of some muscle elements entering into cartilage formation, the question of pluripotency of the blastemal cells remains cloudy. It is entirely possible that the changes termed "dedifferentiation" associated with tissue cells entering the blastema may not represent a complete change back to a primitive undetermined state. Rather, it is likely that the blastemal cell alters its phenotype and capacity for cell division without enlarging its developmental potential. Finally, it should be pointed out that the capacity of muscle cells to contribute to cartilage is at least consistent with the observation involving the regeneration of the distal portion of the humerus from a stump containing no humerus at the time of amputation.

Cell Proliferation and Nucleic Acid Synthesis During Regeneration

Cell proliferation closely follows the pattern of DNA synthesis in the regenerate. The peak of proliferative activity is found proximal to the tip of the blastema in both larval and adult salamander limb regeneration. There is little or no DNA synthesis in the apical epithelium of the paddle stage or in the epithelium covering the phalanges during the finger-bud stage. The distal epithelium continues to be maintained by cells derived from epithelium proximal to the blastema. These cells migrate distally following mitosis. During the paddle stage, those mesenchymal cells engaged in cartilage formation are at first the most actively proliferating cells of the blastemas proper. As development proceeds, the differentiating cartilage cells gradually decreases DNA synthesis. By the finger-bud stage, the bulk of DNA synthesis occurs only along the lateral margins of the developing long bones and bones of the hand.

About 50 per cent of the blastemal cells are synthesizing DNA at the cone stage of regeneration. Virtually all the cells are synthesizing RNA. The apical epidermal cells synthesize RNA, but not DNA. By the paddle and finger-bud stages, as DNA synthesis among blastemal cells is decreasing, virtually all cells still synthesize RNA, including the epidermal cells. As muscle and cartilage cells begin differentiation, some RNA synthesis continues in the myoblasts, while very little RNA is synthesized by chondroblasts during the period of formation of the cartilaginous matrix material. It appears likely that chondroblasts may utilize RNA synthesized in the earlier precartilage period during the period of cartilage deposition. Following the acquisition of a cartilage matrix, the chondrocyte enters a metabolically quiescent phase.

REGENERATION IN ADULT ANURANS

The Role of Continuous Injury

Recovery of regenerative ability in postmetamorphic frogs seems to be associated with tissue trauma and prevention of would healing. Methods include extensive pricking of the amputation surface with a niddle, as well as traumatization using hypertonic salt solutions. Wound healing in adult anurans is accomplished by movement of the epidermis and the dermis over the wound surface. In contrast, only the epidermis covers the wound in urodeles. The anuran response typically continues with the production of connective tissue by the dermal elements and with scarring of the wound. Continued injury in some was interferes with dermalization and permits a variable amount of regeneration to take place. Similar results have been obtained using implanted batteries and the application of a continuous direct current through fine wires to the amputation area. Superficially this might seem to simulate the action of nerves. Regenerates sometimes result from such irritation. Implantation of adrenal glands also prevents dermalization (see below : "Regeneration in mammals").

The Influence of Larval Epidermis

Interestingly, when larval skin is placed over the wound in adult frogs, regeneration becomes possible. Other investigations have implicated the wound epithelium as an important factor in initiating regeneration. For example, head skin inhibits regeneration when placed over the wound in some *Ambystoma* larvae. In these cases, a typical migration of epithelium occurs, but the blastemal cells fail to aggregate and form a regeneration bud.

Other Vertebrates

Studies of successful regeneration in adult frogs have led to repeated attempts to apply the same irritation techniques or nerve implantations (nervous tissue in general seems to promote regeneration) to higher forms which are also without natural regenerative capacities. Particularly this approach has been applied in studies of mammalian regeneration. The limited success to date is described below.

STUDIES ON REGENERATION IN MAMMALS

Limitations in Mammals

Attempts have been made to uncover regeneration potential in mammals. Normally, the regenerative powers in this group are limited to the liver, to replacement phenomena (e.g., the uterine endometrium), and to wound healing. Early experiments dealt with young mammals, which, by analogy with the anuran tadpole, might be expected to show a greater regenerative response than the adult. However, amputated forelimbs of 14-day-old rats *in utero* were found not to regenerate. Studies of wound healing in the toes of young mice did lead to a means of altering the course of wound healing and the production of the dermal scar in these animals. Instead of the softer regeneration blastema, mammalian amputations normally result in wound closure and the production of a fibrous and elastic connective tissue scar formed by the dermal elements of the wound. It was found that the addition of large amounts of ACTH and adrenocorticosteroids, altered the dermalization in the sense that the scarring was delayed but no blastema was formed. Nerve augmentation studies in rat limbs and the treatment of amputated stumps in rats with trypsin, to imitate the early protelytic activity encountered in the regenerating amphibian limbs, were also without result.

The Opossum

The most immature mammal accessible for experimentation is the newborn opossum. This marsupial is found in the mother's pouch in a very immature stage, measuring somewhat less than one-inch long. The hindlimb are very immature though the forelimbs, used to journey from the vagna to the marsupium, are more well-developed.

Amputation of the opossum limb. Amputation of forelimb or hindlimb as soon as 24 hours after birth did not give rise to replacement of any of the lost portions of either limb. However, considerable dedifferentiation of cartilage and bone were noted, as was a thickening of the epidermis over the wound. It appeared that the regenerative response, if any, was aborted soon after it was initiated. The results were considered more encouraging than those previously obtained, and further studies involving nerve augmentation were undertaken. The more immature hindlimb seemed more responsive and was chosen as the system for future experimentation. Even so, the hindlimb contains cartilagenous replicas of the phalanges at the time of birth, so all the parts of the future hindlimb are represented in an immature state.

Nerve tissue implants. Using implants of both opossum and frog nervous tissue (brain), regenerative responses in opossum hindlimbs were observed. The more normal structures were derived from those limbs implanted with homologous brain tissue, though a response was noted to frog brain. With opossum brain implants, a footlike structure appeared after 18 days and contained indications of the fourth and fifth toes. After 32 days, a heteromorphic foot bearing the basal portions of three toes was formed. Replacement of the toes after amputation through the tibia is considered a case of regeneration and no embryonic regulation in these young mammals. No response was observed to implanted adrenal tissue, which had been used with some success in adult anuran regeneration. In no case was there an immune response to the foreign tissue implants; this is consistent with the observation that the newborn opossum is immature with respect to the immune response. Lymphocytes are lacking at birth and the thymus is in the earliest stages of derivation from the epithelium of the third pharyngeal pouch.

Again, the nature of the presumed neurotrophic response is unknown. The limbs of newborn opossums contain nerves of their own, but as yet most of the motor fibers of the hindlimbs are immature, unmyelinated fibers.

2

Regeneration and Pattern Formation

Positional Information

This volume attests a renewal of deep interest in the problem of pattern formation in developmental biology. In large measure, the resurgence of enthusiasm coincides with Wolpert's reformulation of this fundamental problem in terms of the concept of positional information. Prior to Wolpert's introduction of the positional information concept, the dominant theory available postulated the existence of developmental fields possessing "prepatterns", non-uniform spatial distributions of biochemical substances in a tissue, whose local peaks would induce the formation of pattern elements such as specific digits, sensillae, or bristles. In contrast to Stern, Wolpert proposed the more abstract idea that cells within a developmental field possess positional information about their locations with respect to the boundaries of the field, through access to an underlying positional co-ordinate system. The behaviour of each cell in the field was assumed to be due to two independent processes. The cell *assesses* its position, then *interprets* its positional information according to the type of cell it is, and forms a specific structural element in the overall pattern. The chief difference between Wolpert's positional information and Stern's prepattern, is that positional information is free of assumptions about the existence of specific morphogen peaks underlying the subsequent differentiation of specific pattern elements. This freedom in one sense makes the theory of positional information less predictive, yet allows for two important possibilities; first, that the positional information system in *all* the developmental fields of one organism are identical, and, more radically, that the positional information system in all organisms or all epimorphic fields is identical.

The general success of the positional information concept led to a search for the co-ordinate system which supplies positional information. At present, polar, spherical, and generalized Cartesian, models have been proposed. Differences among these alternatives are far from trivial. Although it is always possible mathematically to transform from one to another co-ordinate system, the "forces" or tissue properties which must be postulated to explain the observed features of pattern regulation differ sharply in the different models. While the morphogens have yet to be found, one task in this area of biology consists of efforts to discover the co-ordinate system and "forces" or requisite cell properties that most simply account for the observations, with the hope that the proper formulation will provide both macroscopic laws at the tissue level, and aid in discovery of the underlying molecular variables.

The Phenomena of Pattern Formation

To assess the relative success of the alternatives that have been proposed, it is necessary to review briefly at least some of the major phenomena of pattern generation and regeneration.

Intercalary Regeneration of Intervening Structures

If an amphibian limb capable of regeneration is transected proximal and distal to the elbow, and the distal wrist fragment grafted to the proximal shoulder stump, cell proliferation forms in the wound area, followed eventually by regeneration of the missing elbow region. This basic phenomenon is fairly ubiquitous and fundamental. Juxtaposition of normally nonadjacent tissues from a single developmental field is generally followed by regeneration, in proper spatial order, of the structures normally lying between the juxtaposed tissue edges. This notion of "betweenness" is necessarily central to any theory of pattern formation.

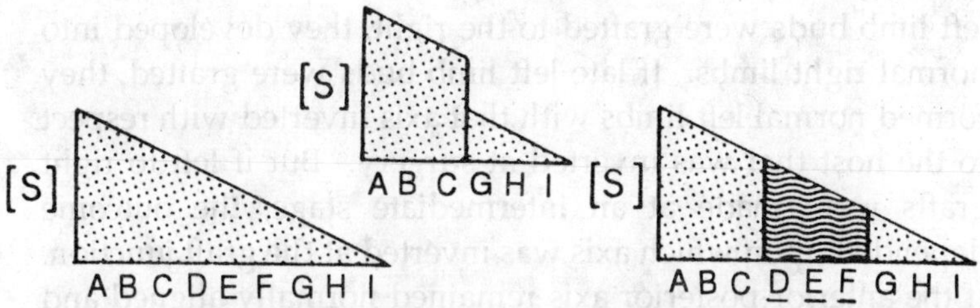

Fig. 2.1. Gradient in concentration of [S] provides proximal/distal positional information in amphibian limbs. Serial threshold levels specify pattern elements A, B, ... 1. Removal of midregion of limb, D, E, F, are grafting, creates discontinuity, stimulating cell proliferation. Diffusive smoothing of gradient discontinuity regenerates missing gradient levels (wavy lines), and structures DEF.

The simplest physical model to account for *"betweenness"* in intercalary regeneration postulates the existence of one or more chemical concentration gradients spanning the tissue, whose concentration levels specify the positional information of cells at each point in the domain. As shown in which a proximal/distal gradient along an amphibian limb is envisioned, surgical removal of the elbow and grafting of wrist to shoulder creates a discontinuity in the gradient, [S], at the graft junction. If one imagines that gradient concentrations remain fixed in the "old" tissue fragments, while diffusion occurs in the new cells of the wound blastema, then simple diffusive averaging of the concentration discontinuity at the graft junction smooths over the discontinuity, recreating all the intervening gradient values in proper spatial order.

Sequential Formation of Positional Axes in Development

In several systems, positional axes appear to be established sequentially during development. In classical experiments, Harrison removed the right forelimb bud of *Ambystoma* and grafted in its stead the left forelimb bud. Such grafts must invert either the anterior-posterior limb axis while keeping donor and host dorsal and ventral axes aligned; or invert dorsal-ventral donor and host axes, while keeping anterior and posterior axes aligned. Harrison found that if very early left limb buds were grafted to the right, they developed into normal right limbs. If late left limb buds were grafted, they formed normal left limbs with that axis inverted with respect to the host that was inverted at surgery. But if left to right grafts were made at an intermediate stage, the outcome depended upon which axis was inverted at the graft junction. If the anterior-posterior axis remained normally aligned and the dorsal-ventral axis was inverted, the donor left limb bud formed a right limb, while if the dorsal-ventral axis remained aligned and the anterior-posterior axis was inverted at the graft junction, the donor left limb bud formed a left limb which remained inverted at the host donor junction. Harrison interpreted his results to imply that the donor anterior-posterior axis becomes autonomously self-sustaining prior to the dorsal-ventral axis. Similar data suggest that amphibian eye axes, as well as limb axes are established sequentially, although the status of data on the eye is in dispute.

Distal Transformation

If an amphibian limb or tail is transected at the elbow, the proximal stump can form a regeneration blastema and regenerate the distal limb. If the digits of the transected distal fragment are implanted into a host flank to establish an adequate blood supply to the distal fragment, the cut surface at the elbow, which initially faced proximally, forms a regeneration blastema and regenerates second distal wrist and hand structures that are mirror symmetrical to the initial implanted distal limb. In these experiments, both the proximal stump and the implanted distal limb fragment regenerate the *same* set of distal limb structures from the cut surfaces at the elbow, identified as a regenerate in the proximal stump and duplicate on the implanted distal limb fragment. The fact that both fragments form distal limb has been called the *"rule of distal transformation"*. Similar results have been found in many insect legs and in the imaginal discs of *Drosophila*.

Supernumerary Limbs

Among the most striking observations in pattern regulation

is the induction of extra or supernumerary limbs following grafting operations. After both the anterior-posterior and dorsal-ventral amphibian limb axes are fixed, transplantation of a left distal limb to a right proximal stump which reverses the anterior-posterior axis of donor relative to host but leaves the dorsal-ventral axis aligned, typically results in formation of two supernumerary limbs at the anterior and posterior margins of the donor host junction. If instead the anterior-posterior axes of host and graft are aligned, but dorsal-ventral axes are inverted, the two supernumeraries emerge from the dorsal and ventral margins of the host donor junction. These supernumerary limbs generally have the handedness of the proximal stump. Similar results have been obtained in transplantation of cockroach limbs. Rotation of a left distal limb by 180° and regrafting to its own proximal stump results in a more variable range of results. After such rotations, the donor may partially rotate back toward its normal alignment; sometimes zero, one, two, or more supernumerary limbs are formed at the graft site, with the same or opposite handedness.

Duplication and Regeneration by Complementary Fragments of a Tissue

Distal transformation by both proximal and distal limb fragments of amphibians is one example of duplication and regeneration by complementary fragments of a developmental field. The phenomenon is common, however, and has been studied in greatest detail in the imaginal discs of *Drosphila melanogaster*.

Drosophila Imaginal Discs

Drosophila melanogaster is a holometabolous insect with egg, larva, pupa, and adult stages. During metamorphosis, the larval ectoderm lyses, and the ectoderm of the adult is formed by the terminal differentiation of special larval organs called *imaginal discs*. In the late third instar larva, each imaginal disc is a two-dimensional sheet of cells forming the surface of a hollow sphere. The columnar cells on one hemisphere form the imaginal disc proper, while the thin squamous cells on the hemisphere form the peripodal membrane which is lost during metamorphosis. Imaginal discs are found as bilaterally symmetric pairs, each destined to form specific left and right regions of the adult ectoderm: the left and right first leg discs form the two prothoracic legs. The two wing-thorax discs form the left and right mesothoraces and wings, etc.

By injecting specific fragments of each disc into host larvae for metamorphosis, it has been possible to construct a fate map of each type of imaginal disc. For example, the fate map of the wing-thorax (hereafter wing) disc, shows that the upper and lower margins of the disc along its longitudinal axis form ventral and dorsal thoracic structures, while the midregion of the disc forms wing structures. During metamorphosis, the wing disc folds along a bent arc running from the anterior to posterior disc margin, and apposes ventral and dorsal thorax areas, ventral and dorsal wing hinge areas, and ventral and dorsal wing blade areas, creating a "bag" that everts through the peripodal membrane. The center of the disc maps to the distal wing tip, while an arcing line from anterior to posterior disc edge through the disc center corresponds to the wing margin.

Grafting experiments like those in amphibian or cockroach limbs are not yet feasible in *Drosophila*. However, closely analogous experiments can be performed by cutting the wing disc into known fragments, and injecting each fragment into the abdomen of a fertilized adult female. In that environment, the disc fragment heals its cut edge, apposing tissue regions which are normally nonadjacent and new cells form in the wound area. Over a week in culture, the mass of a disc fragment typically doubles. After culture in adult abdomens, fragments may be recovered and injected into host larvae for metamorphosis, then recovered from the emerged adult host. By comparison of the patterns of hairs, sensillae and bristles which form when a known disc subfragment is injected directly into larvae for immediate metamorphosis, it is possible to characterize the pattern regulation which occurs in the cultured fragment.

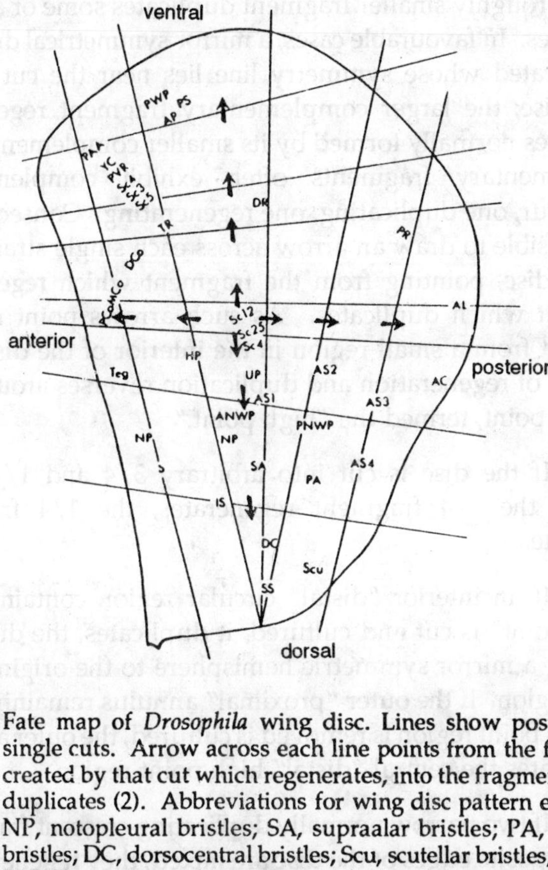

Fig. 2.2. Fate map of *Drosophila* wing disc. Lines show positions of single cuts. Arrow across each line points from the fragment created by that cut which regenerates, into the fragment which duplicates (2). Abbreviations for wing disc pattern elements: NP, notopleural bristles; SA, supraalar bristles; PA, postalar bristles; DC, dorsocentral bristles; Scu, scutellar bristles; ANWP, PNWP, anterior and posterior notal wing processes; Teg, tegula; HP, humeral plate; UP, unnamed plate; AS1-4, first to fourth axillary sclerites Pco, Mco, Dco, proximal, medial, and distal costa; TR, triple bristle row (anterior wing margin); Dr, double bristle row (distal wing margin); PR, posterior row of hairs; Sc4, Sc25, Sc12, groups of sensilla campaniformia on the proximal dorsal radisu; AL, alar lobe AC, axillary cord; PAA, prealar apophysis; YC, yellow club; PVR, proximal ventral radius; PWP, pleural wing process; PS, pleural sclerite; AP, axillary pouch; Sc4, Sc3, Sc5, groups of sensilla campaniformia on the proximal ventral radius; anterior, posterior, ventral and dorsal disc margins.

The following are the dominant results :

1. If the wing disc is cut in two fragments by a straight cut, the roughly smaller fragment duplicates some or all of its structures. In favourable cases, a mirror symmetrical duplicate is generated whose symmetry line lies near the cut on the wing disc; the larger complementary fragment regenerates structures normally formed by its smaller complement. Thus complementary fragments often exhibit complementary behaviour, one duplicating, one regenerating. Consequently, it is possible to draw an arrow across each single straight cut on the disc, pointing from the fragment which regenerates into that which duplicates. As such arrows point radially outward from a small region in the interior of the disc. The polarity of regeneration and duplication reverses around this interior point, termed the "high point."

2. If the disc is cut into arbitrary 3/4 and 1/4 "pie" sectors, the 3/4 fragment regenerates, the 1/4 fragment duplicates.

3. If an interior "distal" circular region containing the "high point" is cut and cultured, it duplicates, the duplicate forming a mirror symmetric hemisphere to the original high point region. If the outer "proximal" annulus remaining after the high point region is removed is cultured, the outer annulus regenerates the central "distal" high point region.

4. If two narrow normally duplicating crescent fragments from opposite edges of the disc are mixed, they regenerate the intervening pattern elements spanning across the disc.

5. It is critical that complementary fragments do *not* strictly duplicate and regenerate. Under different experimental conditions and with varying frequencies, normally duplicating fragments can regenerate most of the pattern elements in the complementary fragment.

6. Anterior-posterior and ventral-dorsal asymmetries exist

in the capacity of narrow duplicating fragments to regenerate. Narrow anterior margin and ventral thorax margin structures have been found by a number of workers to regenerate extensively, while narrow posterior and dorsal margin fragments do not. The results in other imaginal discs are fundamentally similar although it should be stressed that an interior "high point" has not been demonstrated in other discs.

7. Mutants causing pattern duplication in different imaginal discs have been analyzed by a number of workers (e.g., Jurgens and Gateff and will be discussed further.

Models of Positional Co-ordinate Systems

The first major advance in predictive use of the positional information hypothesis lay in the formulation of the polar co-ordinate, or "clockface" model for pattern regulation in epimorphic fields by French *et al.* and Bryant. The initial model was based on some of the results described in amphibian limbs, cockroach limbs, and imaginal discs, and is well illustrated by application to the wing disc of *Drosophila*. The existence of a "high point" in the central (distal) region of the disc about which the direction of regeneration reverses suggested that cells might measure their distance in the tissue from this special point. This raised the possibility that the position of cells is specified in a polar co-ordinate system with the high point as its origin. Since the wing disc is a two-dimensional surface, an azimuthal angle must be measured. Where an angle specified by a single scalar variable, that variable would necessarily be discontinuous along some radial line in the tissue; for example along the radial line ø2π=0. But Bryant found that any 1/4 pie-wedge fragment of the wing disc duplicates, while its 3/4 complement regenerates. If an azimuthal discontinuity were present, the • wedge fragment containing it should behave differently from the rest, and regenerate. Failure to find evidence of an angular discontinuity implies that, if cells measure an angle, they do so seamlessly. Therefore the model proposes that cells measure radial distance from a distal "high point" origin, and an angle seamlessly modulo 2π.

In order to account for the bulk of the data on epimorphic regeneration and duplication, the polar co-ordinate model initially proposed two rules of intercalary regeneration, and a third special rule for distal transformation.

Rule 1. If cells having different radial values are apposed, cell proliferation will be stimulated, and the missing intervening radial values will be restored back to a resting radial gradient, then proliferation will cease.

Rule 2a. If cells having different angular values are apposed, cell proliferation will be stimulated and the missing angular values will be intercalated back to a resting angular gradient.

Rule 2b. Since two angular arcs around a 2π circle of values join any two juxtaposed angular values, a choice rule is needed. The simplest postulates that angular intercalation occurs along the *shorter* arc.

Special Rule 3, the complete circle rule. If a complete circle of angular values at a proximal radial level is exposed, distal regeneration occurs. The special nature of Rule 3 will be discussed later.

The two intercalation Rules 1 and 2a, b, suffice to explain major features of epimorphic pattern regulation. For example, the classical example of grafting an amphibian hand to shoulder with regeneration of the missing elbow region is explained by Rule 1. Proximal radial values in the shoulder apposed to distal radial values in the hand lead to intercalation of the intervening radial values specifying elbow. Duplication and regeneration by complementary fragments of the wing disc is explained by Rules 2a, b. A single, straight cut on the wing disc, yielding a narrow anterior fragment and a large posterior fragment. During culture, the narrow anterior fragment folds over, apposing the cut edge such that the ventral and dorsal thoracic regions heal together, ventral and dorsal wing blade regions also heal together. This healing juxtaposes cells with similar radial values, but discordant angular values. This discontinuity stimulates cell proliferation, and smoothing of the angular discontinuity along the shorter angular arc. Since this shorter angular arc is the arc present in the original narrow anterior fragment, the positional values in the new cells form a mirror symmetrical duplicate of those in the original anterior fragment, and the fragment duplicates.

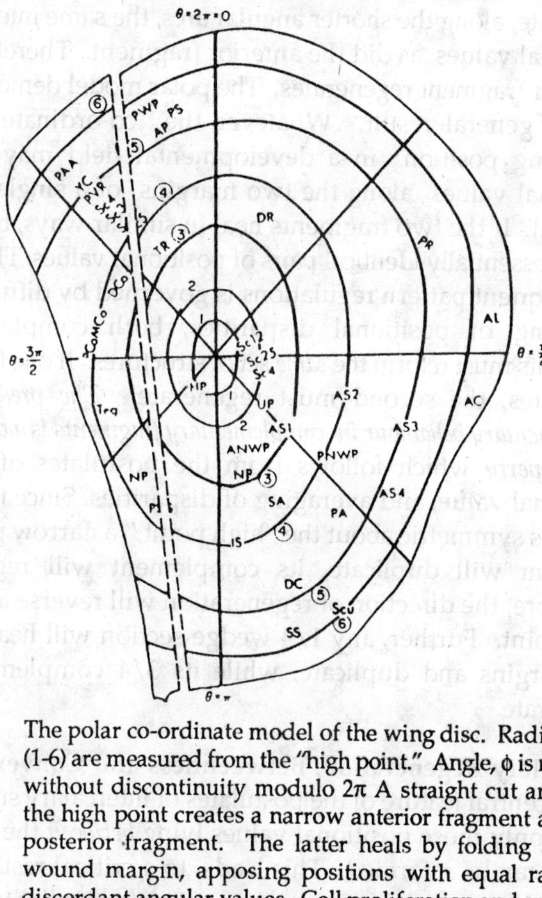

Fig. 2.3. The polar co-ordinate model of the wing disc. Radial values (1-6) are measured from the "high point." Angle, ϕ is measured without discontinuity modulo 2π A straight cut anterior to the high point creates a narrow anterior fragment and wide posterior fragment. The latter heals by folding along its wound margin, apposing positions with equal radial but discordant angular values. Cell proliferation and smoothing of angular discontinuities along the short arc between apposed values leads to regeneration by the posterior fragment. The narrow anterior fragment folds and heals along its wound margin, juxtaposing the same positional values as did the posterior fragment. Short arc intercalation leads to duplication by the anterior fragment.

The positional values present along the cut margin of the large posterior fragment are identical to those along the cut margin of the narrow anterior fragment. If the large fragment wound heals in a similar way, the same pairs of positional values must be apposed in the posterior fragment as in the anterior fragment. Therefore, the posterior fragment must intercalate, along the shorter angular arcs, the same intervening positional values as did the anterior fragment. Therefore, the posterior fragment regenerates. The polar model demonstrates a more general result. Whatever the co-ordinate system specifying position in a developmental field may be, the positional values along the two margins of a single cut are identical. If the two fragments heal in similar ways, both will appose essentially identical pairs of positional values. Therefore, if subsequent pattern regulations is governed by diffusivelike averaging of positional disparities, both complementary fragments must reform the *same* set of structures. If one fragment duplicates, the second must regenerate. *The prediction of complementary behaviour in complementary fragments is a coordinate free property* which follows from the postulates of graded positional value, and averaging of disparities. Since the polar model is symmetric about the "high point," a narrow posterior fragment will duplicate, its complement will regenerate. Therefore, the direction of regeneration will reverse about the high point. Further, any 1/4 wedge section will heal its two cut margins and duplicate, while its 3/4 complement will regenerate.

Intercalary Regeneration, Betweenness and Convex Sets

A central feature of the postulates of intercalary smoothing is that only those positional values lying *between* the apposed values can be reformed. This leads to a critical restriction in the predictive consequences of any given co-ordinate system, since it implies that diffusivelike smoothing can *only* recreate positional values lying in the *convex set* bounded by the positional values in the apposed tissue edges. This restriction, in turn, implies that different co-ordinate systems may demand different special cellular behaviours beyond simple diffusivelike smoothing to account for the data.

The concept of a convex set, and the limitations it imposes can be brought out in the polar model. Radial positions can be visualized without loss of generality as a radially symmetric gradient whose peak is at the distal "high point." I show a wing disc from which the distal high point region has been removed, thus removing the radial gradient "peak." In the remaining outer proximal annulus, only lower values of the radial gradient are present. Therefore, no juxtaposition of tissue edges in the proximal annulus can lead to "diffusivelike" filling in of the missing high-point radial peak. In a polar model, if the region containing the origin is removed, that region does not lie "between" the positional values in the proximal annulus. That is, the region around the origin is not in the *convex set* of all those positional values derivable by averaging *any* pairs of positional values present along the cut margin of the proximal annulus.

The implication of this feature of any polar co-ordinate model is that distal regeneration by a proximal fragment cannot be obtained by diffusivelike averaging of positional discontinuities, and some special rule is needed. In the initial formulation of the polar co-ordinate model, special Rule 3, the complete circle rule was proposed. According to this rule, exposure of a complete circle of angular values at a proximal level leads to regeneration of distal radial values. With the assumption of this rule, the model accounts for the capacity of a truncated amphibian limb to undergo distal transformation, and regenerate distal structures. The same postulate accounts for distal regeneration by a proximal wing disc annulus. Equally, Rule 3 accounts for duplication of the cultured wing disc high point region from the exposed circle of angular values at a proximal radial level, and similarly explains duplication from the proximal cut surface by a newt hand whose digits are implanted into a host flank. Finally, Rule 3 accounts for the striking observation that grafting a left hand to a right stump yields two supernumerary limbs at the position of maximal discord of angular values. Such a graft creates two complete circles of angular values at the radial level of the graft. These undergo distal transformation and yield two supernumeraries, with the handedness of the host. The successes of the polar model are considerable, and have led to testing several features of the model. I turn now to an evaluation of weaknesses in this mode.

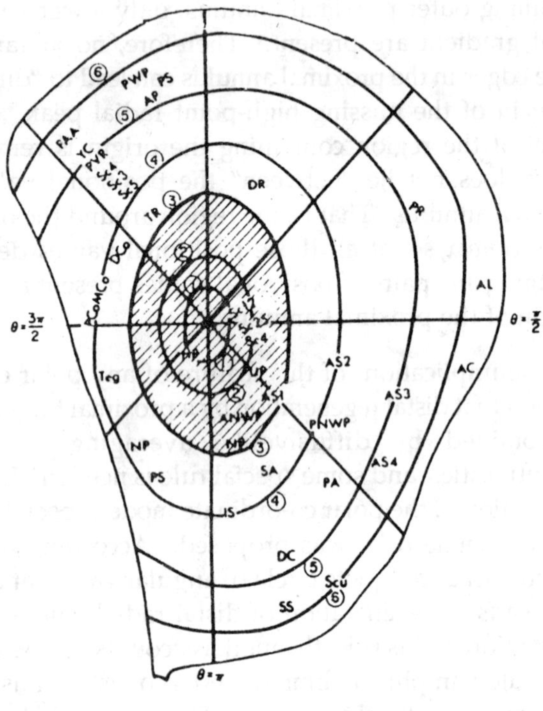

Fig. 2.4. Polar co-ordinate system of wing disc, with distal wing "high point" region removed. No parts of proximal annulus have missing distal radial values. Therefore, no tissue contact and positional averaging can intercalate the missing distal radial values.

First, it should be stressed that special Rule 3, or any modified form of Rule 3, is formally equivalent to postulating a mechanism beyond diffusivelike averaging of positional values, to regenerate a missing radial gradient "peak." For example, in the current version of the polar co-ordinate model, its authors propose that short arc intercalation itself generates distal regeneration. While the postulate of a special mechanism to regenerate distal radial values is not a flaw, its status should be made explicit. As shown next, distal regeneration does not require special rules in other co-ordinative systems. If special mechanisms exist to recreate missing gradient values, then those mechanisms are of central importance. They are likely to play a role in the initial *establishment* of positional gradients as well as subsequent pattern regulation. The forms of special mechanisms which are suggested depend on the choice of co-ordinate system. Thus different co-ordinate systems suggest that different cellular properties beyond diffusive averaging are required to account for the data on pattern formation.

Second, as noted by Winfree, in the vicinity of the origin, the gradient of the angular variable in the tissue becomes infinitely steep, since the origin is a singularity in the angular map on the plane. If cell proliferation is proposed to occur proportional to the gradient in radial or angular values, then proliferation near the origin should be unbounded. It is not. As Winfree points out, this implies a lack of independence of radial and angular gradients near the origin. Third, anterior-posterior and dorsal-ventral developmental axes appear to be established sequentially. This is not deducible, hence not explained, by the proposal that position is measured in polar co-ordinates. For example, if a radial axis is established first, the amphibian limb bud should be angularly symmetric in transplantation between the time when the first axis and second axis are formed, in contravention to Harrison's classical observation. If angular values are established first, then reversal of anterior-posterior or dorsal-ventral areas should yield similar, not dissimilar results, at each stage, but do not.

Fourth, a considerable body of work testing the complete circle rule has shown it to be false in imaginal tissue, cockroach limbs, and amphibian limbs (18, 26, 28). It now appears that distal regeneration is roughly proportional to the proximal angular arc present. Confirmation of the complete circle rule would have been striking evidence in favour of a polar model. Its disconfirmation does not rule out such a model, but alternative co-ordinate systems have the deductive consequence that distal regeneration should be proportional to the proximal arc; hence the data disconfirming the complete circle rule leave a polar model with an *ad hoc* postulate as Rule 3, while alternatives deduce, hence explain, this phenomenon.

Fifth, accumulating evidence shows that normally duplicating imaginal disc fragments can with variable frequency regenerate some, most, or perhaps all of the complementary domain. Such results are now known in the wing, haltere, leg, and genital discs. No convincing comparable data are yet available on amphibian or cockroach limbs or other epimorphic systems, although *divergent regeneration* from symmetrical half limbs in amphibians might be due to regeneration by normally duplicating fragments. The polar co-ordinate model is constructed such that intercalation causes complementary fragments strictly to duplicate and regenerate. The model is unable to account for regeneration by normally duplicating disc margin fragments by simple positional averaging. In fact, no proposal co-ordinate system, coupled with the assumptions of simple diffusive averaging of apposed positional values, can account for extensive regeneration by both the normally duplicating and normally regenerating fragments of a tissue. All co-ordinate systems share the property that both complementary fragments have identical positional values along their cut margins, hence if both fragments heal in similar ways, apposing similar pairs of positional values, both fragments must reform the same set of structures lying in the convex set bounded by the apposed positional values. Evidence that each complementary disc fragment can regenerate large domains of the other fragment is evidence that mechanisms beyond simple diffusive smoothing must operate during imaginal disc pattern regulation. The polar co-ordinate model can account for this phenomenon by weakening the short arc intercalation rule to allow long arc intercalation. Formally, this is equivalent to an active mechanism to regenerate angular values not in the convex set reached by diffusive smoothing.

Sixth, mutants in *Drosophila* are available which can produce completely duplicated legs. In order for a disc fragment to produce a complete duplicate appendage by apposing cut surfaces of the fragment and positional smoothing, the apposed surfaces would have to form a convex set containing the *entire* tissue image. Examination of the polar co-ordinate model shows no single fragment has this property. Within the frame of the polar model, a *partial* disc fragment can produce a mirror symmetric duplicate, but no fragment of a normal disc can produce both a normal leg, and a complete duplicate appendage.

Transverse Gradients, or Modified Cartesian Co-ordinate System

Recently several workers have independently suggested a modified form of a Cartesian coordinate model to account for the data on epimorphic pattern regulation. Suggestions by Winfree and myself are nearly identical. The *Drosophila* wing disc with roughly orthogonal monotonic anterior-posterior and ventral-dorsal gradients of two chemicals, X and Y. Lines of constant concentration are bowed outward on the disc, symmetrically about the "high point." The "image" of this cross gradient system in a "tissue specificity space," i.e., the XY morphogen space. In this "image," lines of constant X or Y chemical concentration, are straight lines, so a convex, bowed line of constant X is a straight line. Similarly, a straight cut on the actual wing disc corresponds to a *concave* line.. The model assumes that a cut wing disc fragment heals its cut margin, apposing nonadjacent positional values, and that simple diffusion smooths discontinuities in X and Y, and fills in the convex set bounded by the apposed XY pairs along the cut margin. As the bowing of lines of constant X and Y concentration on the disc implies that diffusive smoothing of X and Y discontinuities in new cells of a large posterior fragment of a single straight cut will fill the shaded convex set and *regenerate* anteriorly to the anterior-most value present along its cut margin. The complementary anterior fragment apposes the same pairs of discordant values, and duplicates to the same anteriormost value.

Symmetrical convex bowing of X and Y concentrations about the high point insures that the direction of regeneration reverses about the high point. A large anterior fragment from

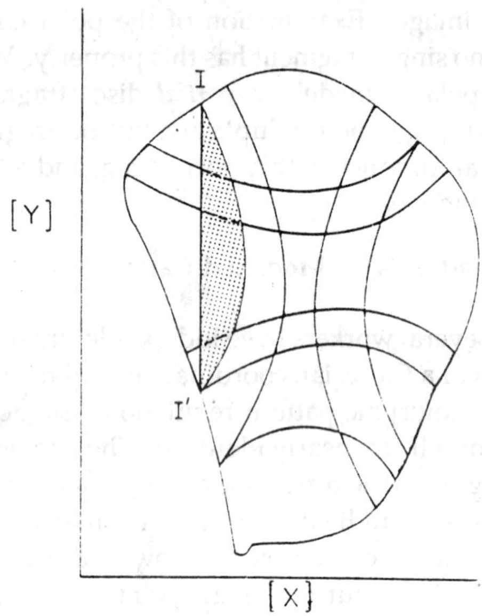

Fig. 2.5. Tissue Specificity Space (33) assigning each point on wing disc a unique value of X and Y. Bowed concave lines are "images" in XY morphogen Tissue Specificity Space of straight cuts on the wing disc like. A single cut from 1 to 1' on the actual disc lies along the concave arc from 1 to 1' in morphogen space. Wound healing apposes the wound margin in the posterior fragment. Juxtaposition creates positional discontinuities that are smoothed by diffusion to fill the convex set, shaded area, bounded by cut margin.

a straight cut will regenerate posteriorly, its posterior complement will duplicate. Similarly, any 1/4 pie-wedge fragment which apposes its two cut margins will duplicate, its 3/4 complement will regenerate. Therefore, a transverse gradient Cartesian model can yield reversal of the direction of regeneration about the high point, without the assumption that the high point is a locus from which cells measure position.

Distal regeneration is a consequence of simple diffusive smoothing in a transverse gradient Cartesian model. As shown, deletion of the high point region leaves a proximal annulus. Wound healing apposes tissue around the circular cut margin, thereby creating discontinuities in the X and Y gradients. Since the deleted high point region lie in the convex set reached by diffusive smoothing from the proximal annulus, the distal high point region is regenerated. Similarly, the high point region itself purse-string closes its circular wound margin, apposing the same pairs of positional values as did the proximal annulus, and therefore duplicates the X and Y high point values in the new cells of the wound area. The same principle predicts that any proximal arc of tissue will undergo limited distal regeneration of structures lying in the convex set of the positional values along the cut margin. Therefore, *limited* distal regeneration proportional to the proximal is present. It has recently been demonstrated in imaginal discs and amputated symmetric half limbs of amphibians and cockroaches, and is a deductive consequence of a Cartesian model.

As a transverse gradient model explains the striking observations that grafts of distal left limbs to proximal right stumps can generate two supernumerary limbs with the handedness of the host. With minor distortion from symmetry, the Cartesian model also has the consequence that 180° graft rotations of the left hand onto the left stump can yield variable numbers of supernumeraries, including two supernumerary limbs of opposite handedness. Finally, unlike the polar model, a transverse gradient Cartesian model is in direct accord with the widespread data suggesting that anterior-posterior axes

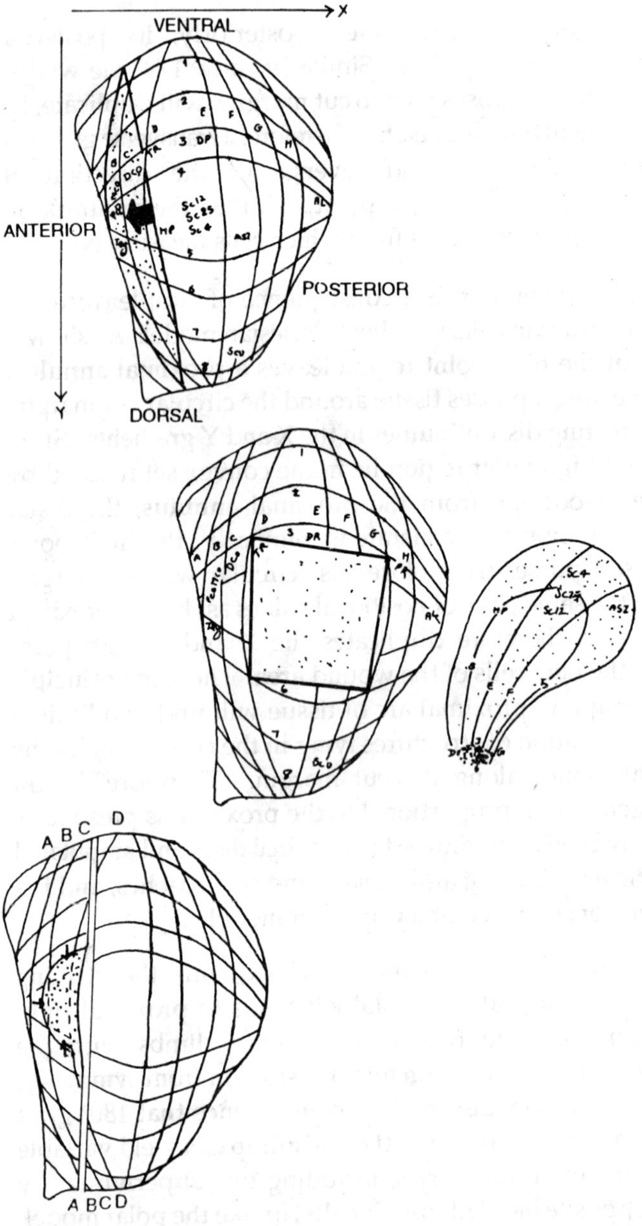
X
VENTRAL
ANTERIOR
POSTERIOR
Y
DORSAL
A B C D
A B C D

Fig. 2.6. (A) A Cartesian co-ordinates system with two monotonic gradients, identical to Figure 5, but showing the concentration of profiles on the actual wing disc itself. The "Y" variable has ascending levels 0-9. The "X" variable has ascending levels A-H. A straight cut creates a large posterior fragment which wound heals by folding in half and juxtaposing opposite ends of the cut margin. For this particular cut, this leads to a discontinuity across the wound in Y values, but not in X values. The convexity in lines of constant concentration of each variable, *isocones*, assures that simple diffusive smoothing of X and Y variables yields regeneration out to the anteriormost X isocone, B, contained along the cut margin of the posterior fragment, stippled area. Similarly, the smaller anterior fragment wound heals the same way and duplicates to the B isocone. The symmetry of convexity in the X and Y isocones on the disc about a central region in the wing disc implies that the polarity of regeneration (heavy arrow) will alter about an apparent "high point" in this region, although cells do not measure position with reference to the high point. Regeneration by a normally duplicating fragment requires an active mechanism beyond simple diffusion to recreate the gradient extremum at the disc margin of the normally regenerating fragment. (B) Removal of a central (distal) region leaves a proximal outer annulus. Wound healing along the cut in the annulus juxtaposes tissues, creating discontinuities in the monotonic X and Y gradients, which are smoothed over by simple diffusivelike processes, leading to distal regeneration by the proximal annulus. The central square containing the high point itself purse-string closes juxtaposing tissues that again create X and Y gradient discontinuities across the wound healed zone. Smoothing of these discontinuities causes the positional values in the central square to be duplicated in the new cells of the wound zone, shown in the granular area. (C) Anterior disc fragment with a distal removed, fills convex set, shaded area, and undergoes partial distal regeneration.

---

are established prior to dorsal-ventral axes. Any transverse gradient model suffers certain defects analogous to those of the polar model in which the "high point" region in the polar model lies outside the convex set of the proximal annulus. Its regeneration requires special mechanisms. In a transverse gradient Cartesian model, the extreme disc margins lie outside the convex sets of their complements. Thus, as the large posterior fragment only regenerates partially. It does not reach the anterior disc margin. Similarly, the narrow anterior fragment duplicates partially. Special mechanisms to recreate X and Y gradient maxima and minima are required to account for the observed capacity of disc fragments to regenerate or duplicate completely.

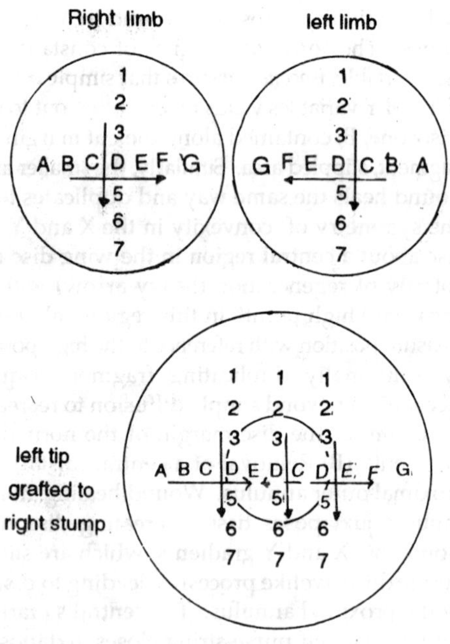

Fig. 2.7. Schematic picture of anterior-posterior and dorsal-ventral Cartesian X, Y co-ordinates for left right limbs, projected onto the plane. The position at the mid values of the two variable, 4, D, corresponds to the distal limb tip. Grafting a left distal limb to a right proximal stump leads to smoothing of discontinuities in X and Y (bold letters, CD and DE) and the formation of two supernumerary distal limb tips with the handedness of the host proximal stump, at the positions of maximal disparity of host and graft axes.

Like the polar model, no Cartesian model can explain the capacity of normally duplicating fragments to regenerate large domains of the complementary disc fragment by diffusive smoothing. Nor can a Cartesian model account by diffusive smoothing alone, for mutants that form complete limbs and complete duplicate limbs, since no subregion of a Cartesian co-ordinated tissue can have the entire tissue image in its convex set. The assumption of a transverse gradient model leads to the suggestion that cells in a tissue have special mechanisms capable of reforming X and Y gradient maxima and minima lying outside the convex sets of cultured fragments. Explicit mechanisms are described next.

Spherical Co-ordinate Model

Russell proposed a spherical co-ordinate model using three orthogonal X, Y and Z gradients to form a Tissue Specificity Space. Position in a tissue is specified by a *solid* angle θ, ϕ corresponding to latitude and longitude angles. Each positional value is a unique ray at a constant X:Y:Z ratio emanating from the origin. That is, each ray is a *line* of equivalent positional value in XYZ space. The longitudinal angle ϕ, is defined by the ratios of X and Y in the equatorial plane, and the latitude angle θ, is defined by the X:Z, or Y:Z ratio. The image in XYZ space of a two dimensional tissue is a two dimensional surface pierced by a set of solid angle rays emanating from the XYZ origin, which itself does not normally lie in the physical tissue.

The spherical model can account for almost all the available data using only the concept of diffusive smoothing. This is most easily visualized in the wing disc by remembering that the disc is topologically a spherical surface, the disc proper, backed by the peripodal membrane. Let this tissue spherical surface be embedded in XEZ space such that each solid angle ray pierces the closed two dimensional surface of the image once, specifying the positional ϕ, θ value at that point. Therefore, the image surrounds the origin, and each ray pierces a unique point on the tissue. If a narrow anterior fragment of the disc is cut, the margin purse string wound heals, and smoothing of X, Y and Z discontinuities in the new cells forms a *second* surface in XYZ space whose edges join those of the original anterior fragment image. Rays from the origin pierce both the new surface and the original anterior hence the narrow fragment duplicates completely. The large posterior fragment wound heals similarly, and forms the same new image surface in XYZ space in the new cells of the wound area. These are pierced by the same rays that pierce the anterior fragment, hence the posterior fragment regenerates completely.

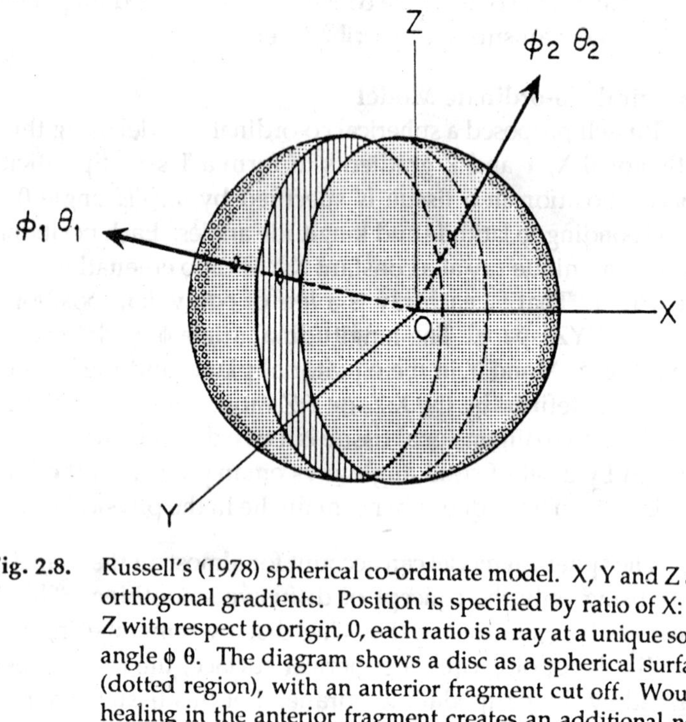

Fig. 2.8. Russell's (1978) spherical co-ordinate model. X, Y and Z are orthogonal gradients. Position is specified by ratio of X: Y : Z with respect to origin, 0, each ratio is a ray at a unique solid angle ϕ θ. The diagram shows a disc as a spherical surface (dotted region), with an anterior fragment cut off. Wound healing in the anterior fragment creates an additional new XYZ surface (parallel lines). ϕ,θ Rays pierce both surfaces, hence the anterior fragment duplicates. The posterior fragment forms the same new XYZ surface and regenerates.

The model is spherically symmetric. Therefore not only do narrow anterior, posterior, dorsal or ventral fragments duplicate while their large complements regenerate, but a distal fragment containing the high point region will purse string close, creating a second image surface in the new cells in the wound area. This second surface will be pierced by the same rays that pierce the high point region, therefore it will duplicate the high point. Equally the proximal annulus will regenerate distally. Finally, this model directly explains distal regeneration proportional to the proximal arc cultured, and the incidence of supernumerary limbs. Despite its strengths, the spherical model suffers weaknesses. First, the spherical model, like the polar model, has a singularity at the north pole high point, where all longitudinal angular values meet. The hypothesis that positional gradients regulate proliferation confronts infinitely steep longitudinal angular gradient at the "high point." More importantly, while the spherical model is very appealing, it is unable to account for all the data by simple diffusive averaging of gradient discontinuities. Like the polar and Cartesian models, diffusion in the spherical model can yield only duplication and regeneration by complementary fragments. It cannot account for regeneration by normally duplicating fragments, nor the existence of mutants which form normal limbs, and complete duplicate limbs. To account for these phenomena, it is necessary to imagine cellular processes capable of recreating missing X, Y or Z maxima and minima not in the convex set of the cultured fragment.

General Critique

The review of the polar, Cartesian, and spherical co-ordinate models leads to several comments.

1. What pattern regulation properties are co-ordinate-free, assuming only the fundamental process of intercalary smoothing of positional discontinuities?

1. Intercalation itself, in one sense is co-ordinate-free occurring in any co-ordinate system with graded scalar variables. Yet it is not entirely free. In the polar model, for example, one might suppose *angular* intercalation occurs faster (or slower) than radial. In the former case, intercalary regeneration would initially spread around annular rings, in the latter, spread radially. In a Cartesian model, assumption of different X and Y intercalation rates leads to yet different directionalities to regeneration.

2. The sequential formation of anterior-posterior or dorsal-ventral axes is not co-ordinate-free.

3. Distal transformation using only positional smoothing is not co-ordinate-free for it does not arise in a polar model. Further, the arguments that distal transformation can be co-ordinate-free implicity assumes embedding the tissue in a two variable co-ordinate system, i.e., in an R^2 space, with no singularities. This is equivalent to assuming *transversal gradients* of scalar variables. It is only coordinate-free in that the gradients need not be orthogonal.

4. Like distal transformation the generation of supernumerary limbs using only intercalation is not co-ordinate-free.

5. Duplication and regeneration by complementary fragments is co-ordinate-free in the restricted sense that both fragments must make the *same* new structures in pattern regulation. However, that any specific fragment should wholly duplicate or wholly regenerate is not co-ordinate-free, as seen in the specific "bowing" of X and Y concentrations in the Cartesian model.

II. The geometries of positional gradients do not identify the co-ordinate system. It might be supposed that determination of actual morphogen gradients in a tissue would suffice to determine the co-ordinate system used by the cells. That this is not so can be seen by reexamining the spherical co-ordinate model. Here the tissue is imbedded in an orthogonal X, Y, Z tissue specificity space. Ratios of these three chemicals specify rays of a solid angle emanating from the origin. A different use of the same three gradients provides a simple molecular interpretation of the polar model. Let the *concentration* of Z specify distance from the high point and the *ratio* of X and Y the azimuthal angle modulo 2π. Each ray in a plane of constant Z, emanating from the Z axis corresponds to a unique angle at a unique Z radial distance from the high point. Simple diffusive smoothing of X, Y and Z, discontinuities recovers the intercalation rules of the polar model. Finally, position might be specified by the actual *concentrations* of X, Y and Z, not their ratios. Thus, the same three orthogonal gradients might be used by the cells in *diverse* co-ordinate systems, with differing predictions about the regulative properties achievable by positional intercalation alone.

III. There is no persuasive evidence of positional singularities. Although the same gradients might subserve different co-ordinate systems, one hope of distinguishing among them is to ask whether singularities exist as expected in the polar and spherical models. The data reviewed offer no persuasive evidence. Not only can "high point" behaviour be generated in a Cartesian model without singularities, but behavioural high points, about which the direction of duplication and regeneration reverse have not been demonstrated in other discs. Indeed, if there is a "high point" in the leg disc, it lies at or near the edge of the upper medial quadrant, and certainly not at the distal tip.

IV. Since no coordinate system can account for regeneration by normally duplicating fragments through intercalation of positional discontinuities alone, further processes must occur in epimorphic pattern regulation. This implies that choice among potential co-ordinate systems cannot be made simply by excluding those, like the polar, that require active processes to explain large fractions of the data.

V. The conceptual framework of positional information itself, and of discussion of alternate co-ordinate systems is inadequate. The general problems are these:

1. Under the postulates of positional information, the pattern which results from *interpretation* of the information is *wholly arbitrary* with respect to the features of the positional field itself. Wolpert's French Flag can switch arbitrarily to a "Wolpert Flag" of Jackson Pollock complexity. No constraints are imposed by the field.

2. Further, even given the assumption of a co-ordinate system, whether polar, spherical, or Cartesian, its relation to the tissue domain is wholly arbitrary and free of the geometry of the tissue itself. For example, in a polar model, we have no grounds whatsoever to think that the high point will be located in the middle of the wing disc and at the upper medial leg disc margin. In short, beyond those properties that flow from smoothing positional discontinuity alone, the hypothesis of positional information is at best weakly predictive.

VI. The observations show that processes beyond smoothing of positional discontinuities in *pre-established* fields are involved in epimorphic pattern regeneration. This suggests that regeneration includes those processes initially involved in pattern generation. While alternative basic premises for considering pattern generation are possible, I suggest the following:

1. Pattern generation is fundamentally self-organizing. While this may not be true in mosaic eggs, the capacity for small duplicating leg disc fragments to regenerate entire legs is but one example of such self-organization.

2. The same example supports a preference for theories which do not postulate special (e.g., source or sink) cells.

3. Intercalation of positional gaps suggests position behaves as though it were spatially averagable. The simplest physical example is diffusion, which suggests that the Laplacian ∇^2 operator expressing such averaging is likely to be fundamental.

The implication of the assumption that diffusivelike averaging of positional variables occurs lies in the fact that such processes naturally couple the size and shape of a tissue to the profiles of gradients which occur in it. The central mathematical concept is an eigenfunction; intuitively this corresponds to a gradient profile having the property that, under the action of diffusion, the profile keeps its *shape*, but may change in amplitude. For example, in a long tube with sealed ends an initial cosine distribution of diffusible stain with maxima and minima at the ends would maintain the cosine distribution but relax to the uniform distribution. Even for modified forms of diffusive-like averaging of positional information, describable by modified operators, the eigenfunctions of such operators provides a natural link between the tissue geometries and the shapes of gradients of positional variables within it. Thematically, the tissue shape controls the profiles of the positional field, which in turn controls tissue geometries and internal inhomogeneities, providing one framework for a self-organizing theory of pattern generation. Further, if gradient profiles are naturally related to tissue shape, it also becomes reasonable to search for natural, rather than arbitrary relations between those profiles, and subsequent interpretation events. For example, "thresholds" might follow contours of constant concentration, rather than being placed arbitrarily.

Consistent with the hypotheses of self-organization, absence of special cells, and diffusivelike averaging, the task is to construct theories able to generate at least two-dimensional positional information. In the next section, I show that the familiar two variable reaction-diffusion model does just this is a particularly natural way.

Sequential Generation of Two Positional Axes and Two-dimensional Positional Information

It is now well-known that model biochemical systems in which synthesis and degradation of chemical species are coupled, and their diffusive transport occurs, can lead to the formation of spatially inhomogeneous gradients of the constituents in the domain. As a particular model, we postulate a single biochemical system of two components, with concentrations X *(r, t)* and Y *(r, t)* at position *r*, time *t*, which are being synthesized and destroyed at rates F (X, Y) and G (X, Y) and are diffusing throughout a tissue. The equations for this system are:

$$\frac{\delta x}{\delta t} = F(X, Y) + Dx\nabla^2 x$$
$$\frac{\delta y}{\delta t} = G(X, Y) + Dy\nabla^2 y \qquad (1)$$

which are chosen to have a spatially homogenous steady state $X_{o,}\ Y_o$ at which $F(X_{o,}\ Y_o) = G(X_{o,}Y_o) = 0$.

Analysis of system (1) begins by linearizing about the spacially homogenous steady state by substituting $X(r,\ t) = X_o + x(r,t)$, $Y(r, t = Y_o + y(r,t)$ and retaining only terms up to first orders in x and y in a Taylor expansion of F(X,Y) and G(X,Y). The resulting linear equations in deviations x and y from the homogenous state are:

$$\frac{dx}{dt} = K_n x + K_{12} y + Dx\,\Delta^2 x \qquad (2)$$
$$\frac{dy}{dt} = K_{21} x + K_{22} y + Dy\,\Delta^2 y$$

The stability of the spatially homogeneous steady state of the linearized equations to spatially inhomogeneous perturbations in the concentration of x or y may be analyzed by evaluating the determinant of the matrix

$$\begin{vmatrix} K_{11} -- k_1{}^2 D_1 -- \lambda & K_{12} \\ K_{21} & K_{22} k_1{}^2 D_2 -- \lambda \end{vmatrix} = 0 \qquad (3)$$

yielding the dispersion relation between the two temporal eigenvalues

$$\frac{\lambda_1{}^{\pm} + {}^1/_2\,[K_{11} + K_{22} -- K_2\,(D_1 + D_2)] \pm}{{}^1/_2 \sqrt{[K_{11} + K_{22} -- k^2(D_1 + D_2)^2 -- 4[k^4 D_1 D_2]^2 -- k^2(D_2 K_{11} + D_1 K_{22}) + K_{21} K_{22} -- K_{21} K_{12}]}} \qquad (4)$$

Here $k_1 =$ k is inversely proportional to the wavelength of the perturbation l_1. For a given wavelength perturbation li, corresponding to a specific value of k_1, if both λ_i + and λ_i have negative real parts, that perturbation decays. In an unbounded domain, all perturbations with wavelengths corresponding to values of k_i for which the dispersion relation has at least one temporal eigenvalue with positive real part, will grow in amplitude in time and create a spatial pattern.

With appropriate constraints on the linearized reaction and diffusion constants (19), the dispersion relation $\lambda = f(k)$ is positive for a restricted range of k, $k_1 > k_i > k_2$, hence for a restricted range of wavelength 1, $l_i < l_i < l_2$ In this case, system (1) acts as an amplifier, selecting and amplifying from thermal noise those wavelength components 1 in the range between l_1 and l_2.

If system (1) exists in a bounded spatial domain and no flux boundary conditions are imposed, then a chemical pattern will be established only if it simultaneously has a wavelength in the range $l_1 = l_2$, and also satisfies the boundary conditions that the gradient have zero component normal to the boundaries. Assuming no flux boundary conditions, the shapes of chemical patterns which can "fit" on to the spatial domain must satisfy the Laplacian diffusion operator, and are an infinite series of specific *eigenfunction* patterns determined by the geometry of the domain. The first several eigenfunctions on a wing disc shape are shown.

Establishment of a First Positional Axis

As the spatial domain in which system acts grows gradually larger, a succession of the discrete eigenfunctions or their superposition will sequentially become established in the domain. For any reaction-diffusion system which acts as an amplifier, perturbations below some minimal wavelength, l_1, will die away. Therefore, if the maximum physical length of the tissue is less than l_1, the spatially homogeneous state is stable. As the tissue grows in length past l_1, the first eigenfunction of the tissue domain will bifurcate from the homogeneous solution. A fundamental theorem of bifurcation theory guarantees that the first pattern will be a stable solution of the full non-linear equations. Therefore, as a wing disc shape gradually enlarges, no chemical pattern forms for lengths below l_1, then eigenfunction 1, grows out of the homogenous state, and forms a monotonic gradient along axis of the tissue.

It has been recognized for sometime that this primary bifurcation provides a means to establish a first axis of positional information in a growing domain and supply one dimension of positional information.

The new question 1 wish to address is: Can a single reaction-diffusion system like (1) establish two transverse axes in a growing tissue?

At first glance, the answer appears to be "No." In the linearized equations of or any similar system, each eigenfunction pattern which forms in the tissue is a spatially inhomogeneous distribution of a *fixed ratio* of the underlying chemical variables x and y, measured as deviations from the homogenous steady state $X_o Y_o$. This property follows from the fact that unstable wavelength l_1 which is amplifying on the tissue, corresponds to a unique scaling factor k_1. Each unique choice of k_1 specifies a particular set of values in matrix having at most one positive temporal eigenvalue l_1; and a unique corresponding eigenvector x_i:y_i, which is a fixed ratio of the linearized deviations of x and y from $X_o Y_o$. This constraint implies that lines of constant concentration of X on the tissue parallel those of Y. Therefore, although two chemical gradients are present in a two-dimensional domain, they supply at most one dimension of positional information. More generally, if N chemical species are in system (1) $N > 2$, all N occur as a fixed ratio when a single temporal eigenvalue λ_i, is positive, hence lines of constant concentration of all N are parallel to one another and supply only one positional axis. Further, if any single eigenfunction pattern, is amplified alone in the tissue through a single positive eigenvalue, the ratio of all N species is again fixed, and at most one positional gradient is established.

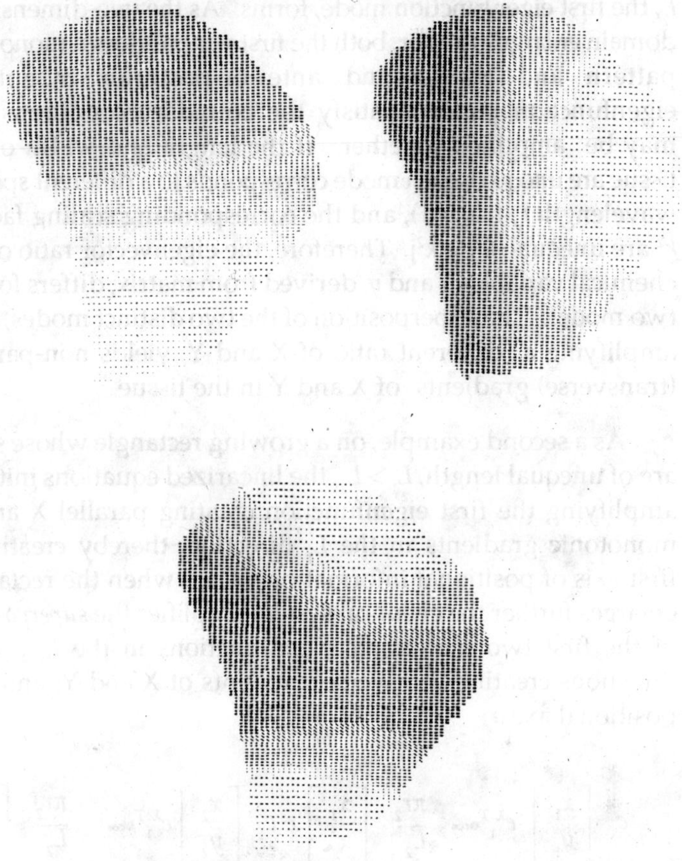

Fig. 2.9. (A-C) First three eigenfunction paterns on wing disc as it enlarges. Concentration coded by visual density (20).

Establishment of a Second Positional Axis.

It has not been generally recognized that the familiar class of two-variable reaction-diffusion systems capable of establishing a first axis can *sequentially* establish transverse axes in growing two-dimensional domains. This, can be exemplified on a wing disc shape. If the range of wavelengths l_1 and l_2 which are amplified by the linearized system (1) is made broad enough, then as the tissue domain enlarges past

l_1, the first eigenfunction mode, forms. As the two-dimensional domain enlarges further, both the first ventral-dorsal monotonic pattern and the second anterior-posterior monotonic eigenfunction pattern, satisfy the boundary conditions and may be amplified together. If the length and width of the tissue are unequal, each mode corresponds to a different specific wavelength $l_{i'}$, l_i ($l_i > l_i$), and the corresponding scaling factors k^2 are unequal, k^2i = k^2j. Therefore, the eigenvector ratio of the chemical variables x and y, derived from matrix, differs for the two modes. The superposition of the two distinct modes, each amplifying a different ratio of X and Y, yields non-parallel (transverse) gradients of X and Y in the tissue.

As a second example, on a growing rectangle whose sides are of unequal length, $L_1 > L_2$, the linearized equations initially amplifying the first eigenfunction creating parallel X and Y monotonic gradients in the L_1 direction, thereby creating a first axis of positional information. Then when the rectangle enlarges further the chemical system amplifies the *superposition* of the first two monotonic eigenfunctions in the L_1 and L_2 directions creating transverse gradients of X and Y, and two positional axes:

$$c_1 \begin{bmatrix} x_1 \\ y_1 \end{bmatrix} e^{\lambda_1 T} \bullet \cos \frac{\pi r_2}{L_2} \begin{bmatrix} x_1 \\ x_1 \end{bmatrix} + c_1 \begin{bmatrix} x_2 \\ y_2 \end{bmatrix} e^{\lambda_1 T} \bullet \cos \frac{\pi r_2}{L_2} \begin{bmatrix} x_2 \\ y_2 \end{bmatrix} \tag{5}$$

The first major property of this class of models, therefore, is the prediction that two transverse positional axes should form *sequentially* as the tissue grows in size.

The linearized model of is of limited value for several reasons. The amplitude of each of the two superimposed modes grows exponentially without bound. Whichever mode grows faster (λ_1 or λ_2) eventually dominates the superposition, therefore a single eigenvector $[x_1 : y_1]$ or $[x_2 : y_2]$ dominates the pattern and the underlying X and Y gradients become parallel. Second, the contributions of the first and second patterns to the superposition are given by arbitrary initial conditions, c_1, c_2. Finally, each eigenfunction pattern can occur in two opposite forms or branches, with maxima replacing minima. Which is established depends entirely on initial conditions. The conjunction of these defects implies that the relative orientation of X and Y gradients on the tissue is not uniquely fixed by tissue geometry in a linear theory.

These defects of a linear model are fully met by a fairly broad class of non-linear reaction-diffusion systems in an *asymmetric* two dimensional tissue such as the wing disc. A number of workers have proposed non-linear models which are known to bound the amplitudes of chemical patterns in one spatial dimension and two dimensions. Generally, the establishment of the first pattern parallels the linear analysis, except that for large amplitudes, the pattern may deviate from the eigenfunction pattern. As the spatial domain enlarges, the second pattern commonly is gradually established through a secondary bifurcation from the primary spatially inhomogeneous monotonic pattern, yielding at each tissue size a fixed mixture or amplitude of distorted first and second modes. Finally, in a spatially asymmetric two-dimensional domain, the cube of the eigenfunction is almost always non-zero. From bifurcation theory this implies that, as the tissue enlarges, only one of the two branches (or forms) of the primary monotonic pattern emerges smoothly from the homogenous state, while the other branch is stable but requires a macroscopic perturbation to be attained. For at least some non-linear systems, only one branch of the subsequent secondary bifurcation emerges smoothly from that first mode. The novel implication is that the entire sequential establishment of a first positional axis, then two transverse positional gradients in two dimensional domains, their shapes and orientations, can be uniquely determined by the changing tissue size and geometry, and the kinetic parameters of the non-linear model.

In Fig. 10a, b, I show the results of simulations of one non-linear model given by :

$$\frac{\delta X}{dt} = -- AX + \frac{BY^6}{1 + Y^6} + D_1 \Delta^2 X \qquad (6)$$

$$\frac{\delta Y}{\delta t} + -- CX + \frac{D(Y6 + 6)}{1+Y^6} + D^2 \Delta^2 X$$

with parameter values specified in the legend. As a wing disc shape enlarges, a first ventral-dorsal monotonic gradient of X and parallel gradient of Y is established and supplies one-dimensional positional information along one positional axis. As the disc enlarges further and the underlying anterior-posterior second mode begins to contribute to the pattern, the X and Y gradients both rotate slowly on the disc, Y faster than X, gradually establishing distinct non-parallel transverse, but not orthogonal, monotonic X and Y gradients spanning the tissue forming a second positional axis. The model sequentially establishes transverse axes and two dimensions of positional values through two gradients whose forms and orientations are initially identical, but gradually come to differ as the tissue enlarges. The positional field is uniquely determined by the shape trajectory of the tissue, and the specific non-linear model.

Novel Predictions

The inherent simplicity and minimality of this two-variable reaction-diffusion mechanism commends it as a serious model for the establishment of transverse positional axes and positional information in developmental fields. It may be that a more complex three variable model coupled with the superposition of three or more modes can yield three transverse gradients which might underly Russell's spherical co-ordinate model or a polar co-ordinate model. The minimal model has a number of predictive consequences.

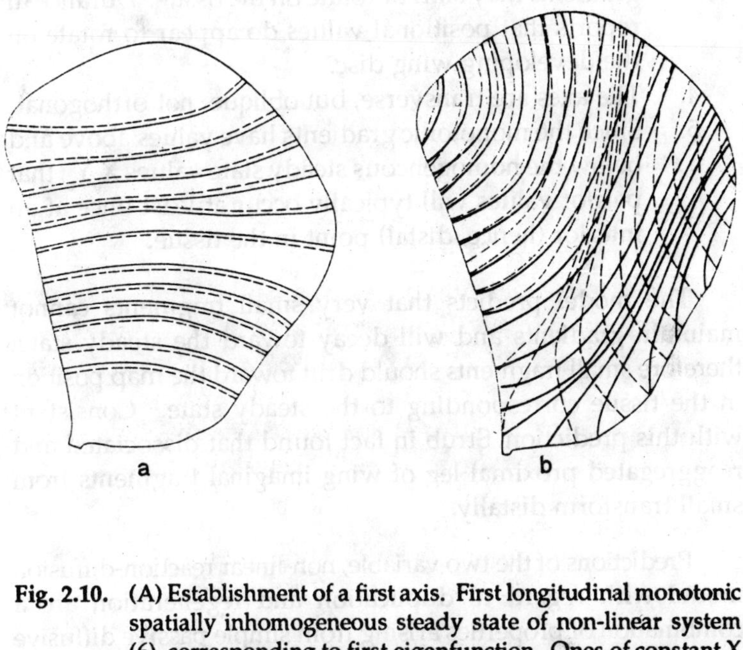

Fig. 2.10. (A) Establishment of a first axis. First longitudinal monotonic spatially inhomogeneous steady state of non-linear system (6), corresponding to first eigenfunction. Ones of constant X concentrations (solid lines) parallel lines of constant Y concentrations (dashed lines). Spacing between lines of constant X concentrations, or of Y concentrations are not strictly proportional to gradient slope, but are used to indicate shapes of X and Y gradients. (B) Establishment of Second Positional Axis. As wing disc grows, second eigenfunction, makes a contribution to mixed non-linear steady state solution. X and Y gradients rotate away from positions in (A), and lines of constant X concentrations (solid lines) separate from lines of constant Y concentrations (dashed lines) creating transverse gradients and a second positional axis. Parameter values used for the non-linear model (6). A = 7.8, B = 15.6; C =1; D = 1.87; b= .2; D_1 = 16; D_2 = 1.

1. Positional axes are established sequentially, as seen in several systems.
2. The longer axis in a field is established before the shorter.

3. During establishment of the second axis, the positional gradients may shift or rotate on the tissue. Duranceau reports that positional values do appear to rotate on the developing wing disc.
4. The axes are transverse, but oblique, not orthogonal.
5. Since the monotonic gradients have values above and below the homogeneous steady state values $X_o Y_o$, that pair of values will typically occur at least once at an interior (hence, distal) point in the tissue.

The model predicts that very small fragments cannot maintain gradients and will decay toward the steady state; therefore small fragments should drift toward the map position in the tissue corresponding to the steady state. Consistent with this prediction, Strub in fact found that dissociated and reaggregated proximal leg of wing imaginal fragments from small transform distally.

Predictions of the two variable, non-linear reaction-diffusion model with regard to duplication and regeneration are a combination of properties arising from simple passive diffusive averaging of positional dicontinuities, and properties arising from the active dynamic and boundary conditions, and are described more fully elsewhere.

It is important to stress that the capacity of simple two-variable reaction-diffusion systems to generate transversal gradients supplying two-dimensional information is not restricted to a special subset of these models, but is rather a robust feature to be expected in essentially all members of this class of models. The general reason this is so is that these models generically amplify a finite *range* of wavelengths, L_1 - l_2. As the tissue in which such a reaction-diffusion system occurs grows larger, or other parameters alter, it must eventually reach a size at which *N* of the longer wavelengths "fit" on the tissue, while *N* + 1 of the shorter wavelengths simultaneously fit. Since both modes satisfy the boundary conditions, both will be amplified and the resulting (non-linear) superposition of gradients will be locally transversal. A very substantial amount of mathematical work will be required to clarify the character of such patterns. I believe this class of models which can couple the shape of a tissue to the shape of the transverse gradients which arise in the tissue, warrants deeper exploration.

Important directions for further elaboration of this class of theories include:

1. Exploration of predictive rather than arbitrary relations of the positional field to differential cell behaviour within it. Examples among these are thresholds as contours of constant concentration, and polarization of cells or alignment of division products by the local slope of the field. One past use of the successive eigenfunction on the wing disc has been to account for the numbers, positions, two-fold symmetries and sequence of wing disc compartmentalization.
2. Analysis of the implications of non-monotonicity in positional gradients. The typical behaviour of pattern generating mechanisms like reaction-diffusion systems with preferred wavelengths, as a tissue grows, is to shift from monotonic to non-monotonic gradients. Positional information then becomes nonunique, with adjacent mirror symmetric domains, and repeated units of pairs of such domains. Maintaining unique positional information might be achieved by several alternatives and may have implications for the emergence of secondary fields in general.

3

A Continuity Principle for Regeneration

Many kinds of living organisms regrow appendages that are crushed or torn off in the mishaps of an active life. Human beings have scarcely any abilities of this sort, a fact which contributes to their curiosity about the mechanisms of regeneration in more resilient organisms. This curiosity runs deeper because regeneration in many ways resembles the initial normal development of an animal's structures. Normal development plus regeneration, collectively called *morphogenesis*, presumably operates by some general rules that we might at least elucidate empirically as a prelude to ferreting out deeper mechanisms. Yet for all the imaginative and neticulous efforts of at least four generations of developmental biologists, few general rules have stood the test of time. If principles of widespread applicability exist they remain tantalizing obscure.

In this situation, the whimsically dubbed *clockface ("clock-phase")* model of regeneration created a stir among developmentalists of both experimental and theoretical bents. Its most fundamental and radical tenet is that positional information is stored in cells as a "phase" or an "angle" analogous to the hour-hand position on a face of a clock, a quantity topologically unlike the conventional scalar variables of physiology. This has several startling implications which have been verified experimentally in several contexts, so it seems to call four some interpretation, perhaps involving a rhythmic or somehow periodic process.

It also turns out that this model, unless adorned with special hypotheses to cover for exceptions, makes other predictions that are not observed. Moreover, a less original interpretation suffices to account for the remarkable regularities that gave the clockface model its first appeal. This relatively prosaic rendering assumes that positional information is encoded, not in a single angle or phase, but in a pair of transverse concentration gradients. This is a completely ordinary idea, but its implications were long overlooked. They include the essential qualitative facts about regeneration in epimorphic fields.

The Clockface Model

The clockface model was contrived as the vehicle for a brilliant resynthesis of observations by Vernon French, Susan Bryant, and Peter Bryant. They draw on experiments with the fruitfly larva's "imaginal disks" (the precursors of the adult fly's various appendages), with the legs of cockroaches, and with amphibian limbs. Their original paper) is a *tour de force* of focused experimentation and organization of data. They come up with two principles which together account for an extraordinary diversity of peculiar experimental results. Let me first remark that what one takes to be "the principles" is presently subject to so much subjective interpretation that there are as many versions as there are people retelling this story I will retell it based on four principles. The reader will want to examine the bibliography for other versions and for more detailed references to the original experimental papers. It should also be noted that the experiments under consideration here do not span the whole diversity of regenerative processes. We focus here on situations in which cells seem to have no intrinsic directional polarity, seem to retain their differentiated identity, and seem to impose on adjacent cells (possibly newly created from dividing populations) to adopt a similar identity.

The main point introduced by French et al. is that each tiny patch of tissue is somehow labeled permanently with an unchanging intensive quantity — a tissue specificity — that behaves like time in a cycle or like a hue or like an angle in that any one of them denotes a point on some abstract circle of states. French et al. (4) placed the digits 1-12 as indicators of local tissue specificity around the circumference of a limb or other developmental field. Many of their diagrams thus resemble the face of a clock. Although this notation gives the model its name, it unfortunately also requires a numerical discontinuity where none is intended biologically. A more apt analogy is provided by the seasonal states of an ecosystem: spring grades into summer grades into fall grades into winter grades into spring with no discontinuities. The numerical analogy is unfortunate, but the principle is clear and tantalizingly suggests something oscillatory, somehow periodic, underlying. Without committing ourselves to any such interpretation, it will still be convenient to accept the "clockface" metaphor and refer to the quantity denoted around its rim as a "phase." Independent of its circlelike topology, the import of a cell's neighbourhood bearing this putative state label is that an an appropriate stage in regeneration, tissue will develop structures (sensory hairs, muscle attachments, colour patches) corresponding to the local label. This is an application of Wolpert's principle that tissue specificity, encoded in a local quantity of unknown origin, is conceptually distinct from the cell's interpretation of that "positional" specificity and from other convert internal states such as "determination."

The following simple rules make use of the putatively *circular* label of cell states, and suffice to systematize a lot of otherwise very perplexing experimental results.

Rule I : Each little patch of cells is labeled with a phase which is part of a smooth phase gradient across the tissue. Thus there exists a smooth map from the tissue to an abstract ring of biochemical specificities. Once established, this map does not change. (Some independent second label, e.g., the proximo-distal level of cell specificity, is implicitly assumed, whereby to distinguish cell types on the two dimensions of an organism's surface.

Rule 2 introduces an exceptional point near which this map is not smooth.

Rule 2: In the normal limb, the phase values are supposed to run one full cycle around the limb axis (the azimuthal direction). If a limb were an open-ended cylindrical surface, this would pose no problem. But besides having an inside, here discretely ignored, a limb has a foot or a hand at the end: its surface resembles the glove with phase increasing full cycle around the open end. Herein lies the tantalizing nucleus of this whole subject. It phase increases smoothly through a cycle around the perimeter of a two-dimensional patch of tissue, let us say that the patch has "winding number" $W = 1$ (or 2, should phase increase through *two* full cycles, etc.). It is an unfortunate fact of geometry that only if the winding number is *zero* can a distinct phase value be attributed to every point inside. To convince yourself of this, try to sketch on the tissue a curve along which phase = 1, and an adjacent curve along which phase = 2, and so forth. All these curves start from the corresponding points along the tissue's boundary, and extend inward. If the winding number of phase along that border is zero, then each curve that enters the patch from one point on the border can again exit the patch at another point where the phase value is the same. But if the winding number is not zero, then that integer number of full sets of phase contours enter and can't get back out. Wherever they converge, two points an arbitrarily small distance apart may have radically different assignments. This is a discontinuity. If $W = 1$ there must be a discontinuity in the assignment of phase values somewhere; the tissue could evade this discontinuity only by placing it in a hole (a region devoid of living, labelled cells). one might suppose that the discontinuity would appear along some "seam" running the length of the leg (the proximo-distal direction) like an International Date Line, but none was found. The discontinuity is apparently more compactly localized than that. It might be an isolated phase "singularity". Alternatively, there could be several isolated phase singularities, e.g., three, of which two have opposite handedness and cancel out.

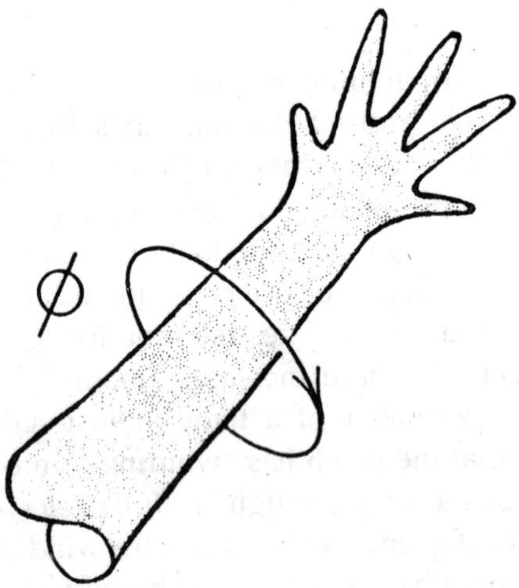

Fig. 3.1. The circular dimension around an appendage is putatively encoded in the cells by a time-independent quantity that increases smoothly from an initial value and returns to it without ever decreasing. For the sake of graphic clarity, several diagrams here use humanlike appendages rather than the more diverse limbs peculiar to the actual experimental subjects; humans, of course, do not regenerate much more than a fingertip.

Rule 3 corresponds to the Rule of Intercalation.

Rule 3: So long as the phase gradient is shallow enough (if adjacent cells are sufficiently similar in phase), cells divide only to replace those that happen to perish. But if cells with normally non-adjacent phases are juxtaposed (e.g., by cutting off a leg, rotating it, and sticking it back on), then those cells begin to divide in earnest. As proliferation continues, the new cells take on phase values intermediate "between" their immediate neighbours. Thus the phase discontinuity is soon bridged through newly regenerated tissue. Proliferation continues until it has restored the initial shallowness of the phase gradient in space.

In some versions of the clockface model, intercalation of intermediate phase values might, without an additional rule, go either of two ways: there are two arcs of phase values "between" any pair of points of a circle. We defer this puzzle until it can be seen in its more general aspect.

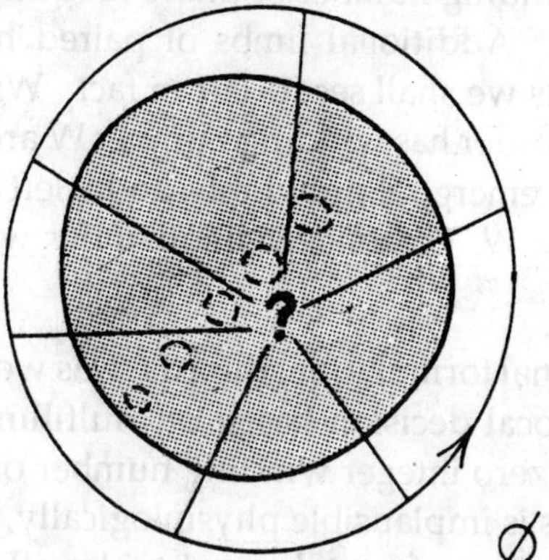

Fig. 3.2. The epidermis is stretched flat. The dashed rings indicate fingernails. From the five points on the abstract circle of phase values contour lines cross the epidermis, joining cells of like value. If the perimeter spans a full cycle, then these contour lines inevitably converge on the borders of an internal hole, or, if there is none, on a region of ambiguous phase (a "phase singularity").

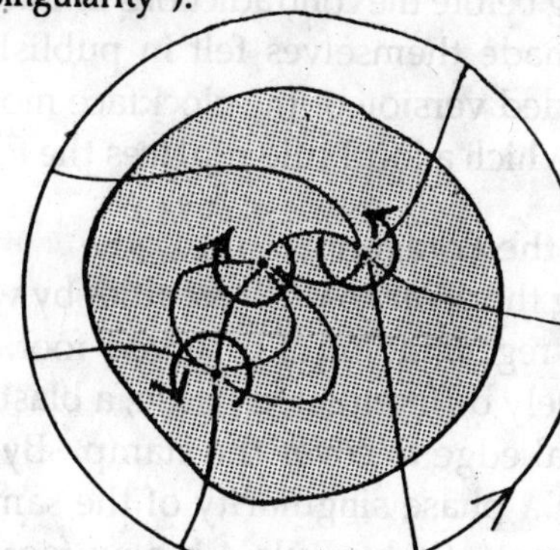

Fig. 3.3. As in but showing that any picture may be supplemented by any number of additional paired singularities of opposite handedness without violating the boundary conditions.

Rule 4 corresponds to the second rule of, the Complete Circle Rule of distal outgrowth.

Rule 4: A new limb will grow out wherever there is a phase singularity. It is left- or right-handed according to whether the winding number of phase around the singularity is + 1 or — 1. Additional limbs of paired handedness are possible, and as we shall see, occur in fact. Within a patch of tissue whose border has winding number W around the phase ring, there will emerge *R* right limbs and *L* left limbs in such a way that *R* - *L* = *W*. For an exposition of this winding number formulation.

In its original formulation, this rule was worked as though cells make a local decision based on fulfillment of a global criterion (non-zero integer winding number on a distant ring of tissue). This is implausible physiologically, and lends itself to the construction of paradoxes, besides. In the approach taken here, emphasizing strictly local principles of cell interaction, something like the Complete Circle Rule emerges as a mathematical artifact, valid in especially simple situations. These difficulties prompted several independent analyses that converged on sub-stantially the same ideas as presented next. It was not long before the contradictions implicit in the Complete Circle Rule made themselves felt in published experiments, too: an amended version of the clockface model subsequently appeared in which a local rule replaces the former global rule.

Examples of the Foregoing

Applying the polar co-ordinate rules by way of illustration, consider the regeneration of a served foot. A disorganized layer of vaguely differentiated tissue, a blastema, grows from the cylindrical edge to cover the stamp. By Rules 1 and 3, it must contain a phase singularity of the same handedness as the original foot. So by Rule 4 a new foot of the same sort replaces the old. In terms of winding numbers, W = 1 (or —1), so in the simplest case *R* = 1 and *L* = 0 (or *L* = 1 and *R* = 0).

Consider a second example in which a right foot is cut off and replaced by a left foot served from the other leg. The cylinder of new tissue proliferating in the junction according to Rule 2 is bounded by two complete circles of phase, as indicated. This two - part border has winding number $W = 2$ around the cylinder of skin it encloses. This is shown by slitting the stippled skin along the line AB and laying it flat : along path ABB 'A'A the phase increases by 0 +1 + 0 + 1 cycle. Thus we expect two additional new right feet to emerge. In animals capable of regeneration, the actual result is indeed two additional right feet. A Chinese woman suffered this very operation in 1973, following piecemeal destruction of both limbs in a railroad accident (15). The present inability of humans to regenerate whole limbs presumably spared embarrassment to all concerned.

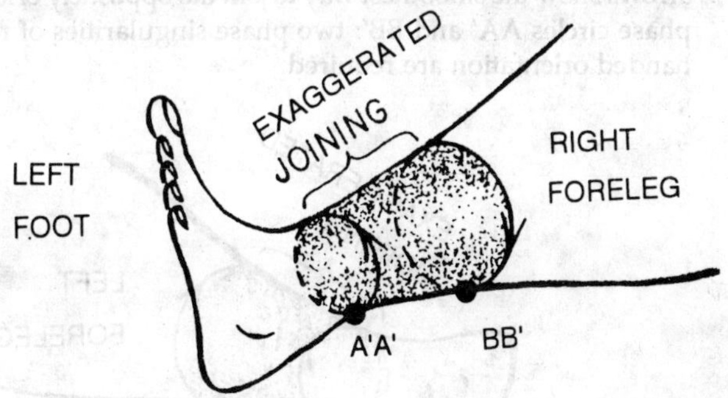

Fig. 3.4. In the stippled cylinder of regenerating and healing tissue, the transition is made from the clockwise orientation of the left-handed graft (foot) to the anticlockwise orientation of the right-handed host (foreleg).

As a last example consider a left foot cut off and replaced with a 180° rotation. By the same argument, we expect no supernumerary limbs, or else any number of left-right pairs. A common result is a right foot and a left foot. The other common result is no new limbs.

An amusing implication of all this, incidentally, is that no organism capable of regeneration in this mode can have an

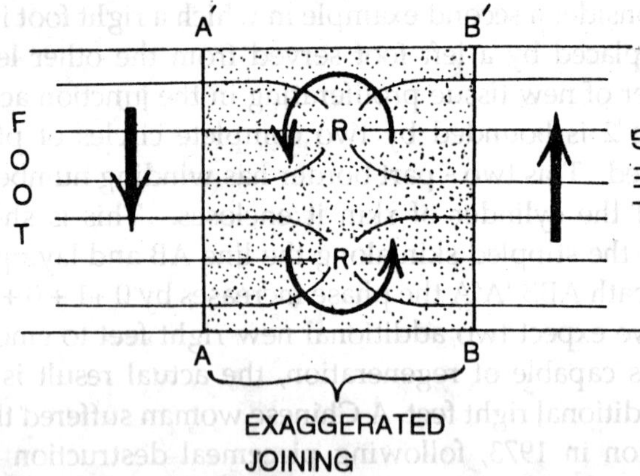

Fig. 3.5. The stippled cylinder is slit open along the line from AA′ to BB′. Lines A′B′ and AB are the same. Contour lines and arrows show the smoothest way to join the oppositely oriented phase circles AA′ and BB′: two phase singularities of right-handed orientation are required.

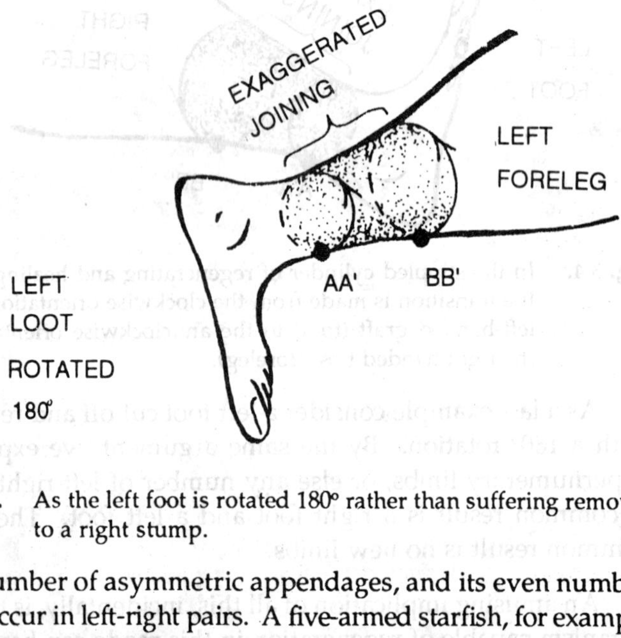

Fig. 3.6. As the left foot is rotated 180° rather than suffering removal to a right stump.

odd number of asymmetric appendages, and its even number must occur in left-right pairs. A five-armed starfish, for example must be a covert tetrapod with a prehensile tail (a bilaterally fused appendage)—as it is in fact, the echinoderms being descended from bilaterally symmetric ancestors.

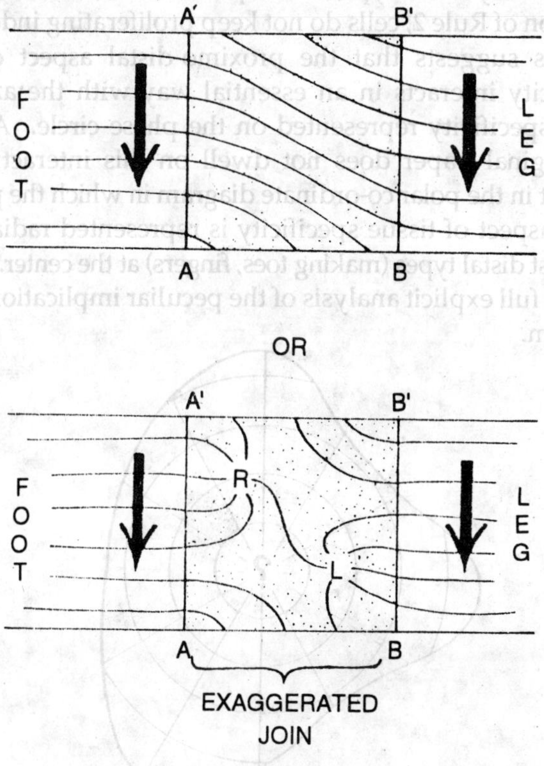

Fig. 3.7. As transition can be made without singularities, by just twisting all contour lines through 180° (top). An alternative construction (bottom) intercalates a superfluous pair of complementary singularities. Both outcomes are common.

Alternative Formulations

Superficially, the data compiled by French et al., would seem to argue forcibly that phase singularities do play a central role in the morphogenesis of higher animals, much as they do in wave-conducting sheets of oscillators such as the social amoebae, the ascomycete fungi, and certain chemical reactions.

This is a distinct possibility that I personally find very exciting, but it does have difficulties. These arise when we ask, as in the other organisms exhibiting phase singularities, what happens at the singularity? There is no abrupt discontinuity in cell type, no unique structure, and in apparent violation of Rule 2; cells do not keep proliferating indefinitely. All this suggests that the proximo-distal aspect of tissue specificity interacts in an essential way with the azimuthal tissue specificity represented on the phase circle. Although the original paper does not dwell on this interaction, it is implicit in the polar co-ordinate diagram in which the proximo-distal aspect of tissue specificity is represented radially with the most distal types (making toes, fingers) at the center. Sibatani gives a full explicit analysis of the peculiar implications of this diagram.

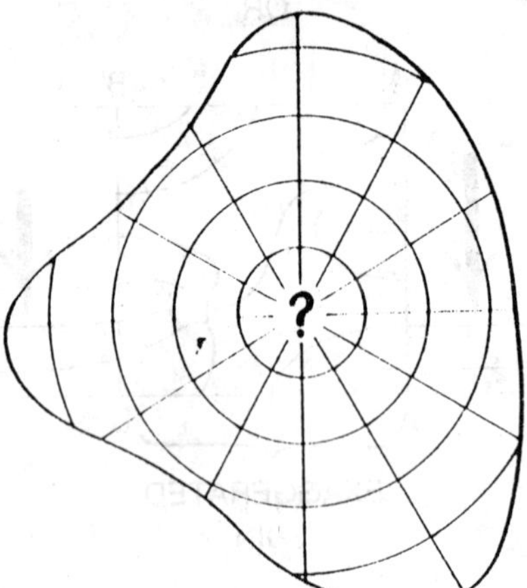

Fig. 3.8. On a piece of generic tissue, level contours of proximo-distalness are drawn as rings concentric to the physical tip. Level contours of a complementary aspect of tissue specificity, locally transverse to the proximo-distal gradient, close in rings. The "?" denotes ambiguity in the phaselike quantity where all values ostensibly coexist without perpetual growth.

The landmark contribution that came with the clockface model consisted primarily of a clear statement of correlated phenomena: There are some astonishing regularities spanning diverse phyla, never before perceived so lucidly. The model itself had less appeal, its own internal contradictions lying so near the surface. The inevitable flurry of counter-experiments and *ad hoc* amendments was well summarized, but the ostensible elegance of 1976 was irretrievable once the subject had been glimpsed from another perspective. At present it appears that the reported phenomenology is best understood as a consequence of little more than the continuity properties of any topologically suitable representation of "positional information" or (to avoid confusion with physical position) "tissue specificity" underlying visible differentiation. The most tantalizing feature of the clockface model was its use of a *circular* co-ordinate for measuring cell types. It cells have two interdependent aspects of tissue specificity then one must ask exactly what motivates the choice of one co-ordinate system over another on this two-dimensional state space. Of course, one would prefer a co-ordinate system natural to the topology of the state space.

An Inverse Fate Map

A simple co-ordinate-free representation might be constructed as follows. Let us imagine a tissue-specificity space (TSS). We endow it with enough dimensions (two) to distinguish cell types on a two-dimensional surface such as the surface of a leg. We depict as a *place* in TSS the latent tissue specificity of a cell or a patch of neighbouring cells. To each region in TSS there corresponds a type of structure that cells in that region will make when they mature. Map the creature, organ, or tissue into TSS. There is nothing new in this procedure. This is only drawing a fate map, inside out as it were: instead of drawing the organism in the real world and writing tissue names on an overlay, we write the tissue names at fixed places in tissue-specificity space and then draw the organism as an overlay, distorted as required to put places on their corresponding names. For example, this "inverse fate map" of any bilaterally symmetric organ is folded so that two patches of cells symmetrically disposed to the left and right of a mirror plane both map to the same place in TSS. The mirror axis maps to a fold line.

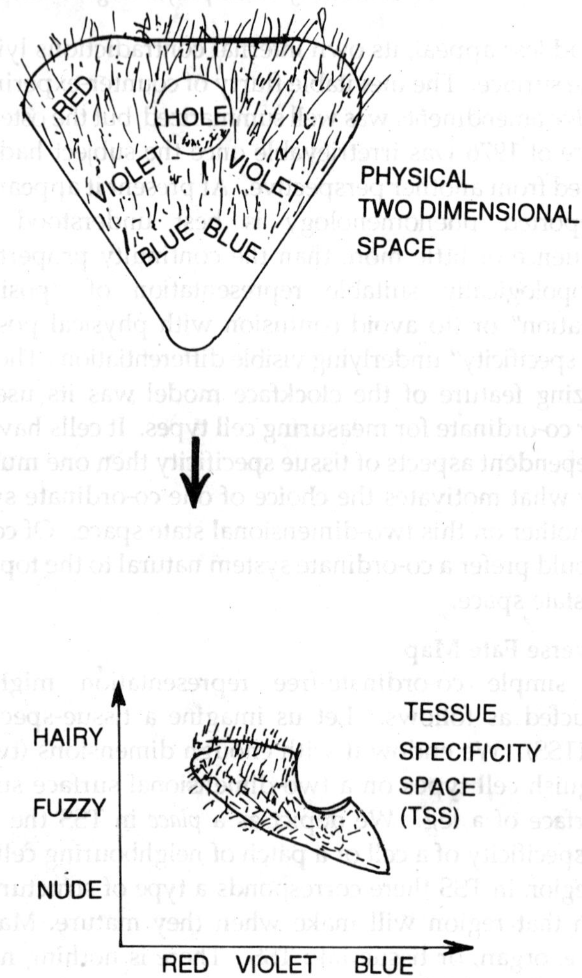

Fig. 3.9. An imaginary two-dimensional organism and its image in tissue specificity space.

Now consider two pieces of an organism, normally not adjacent in the intact, mature stable organism. These appear in TSS as two islands of tissue. If they are now physically juxtaposed, tissue specificities are not initially affected, so fine lines must be used to connect the cells that are quite separate in TSS though physically adjacent along the surface of contact.

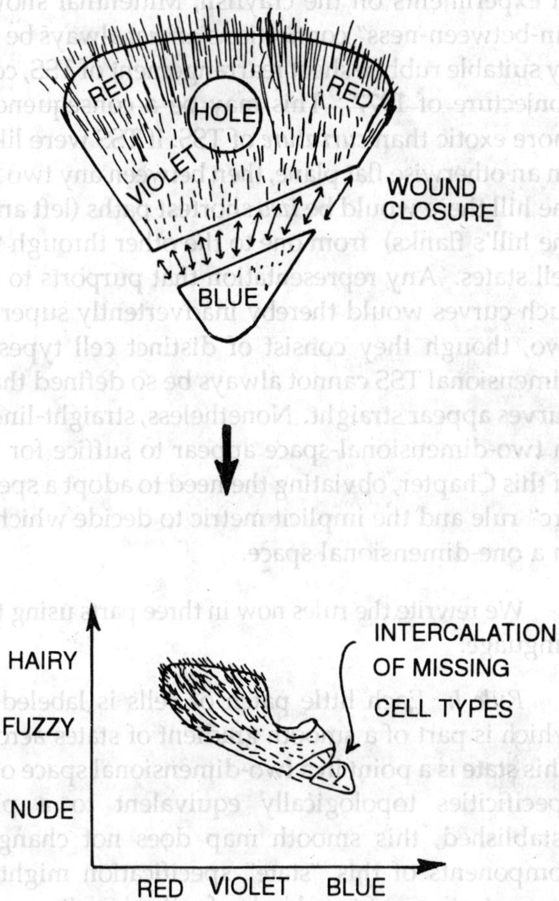

Fig. 3.10. A strip of tissue is removed from the organism and its image without affecting adjacent tissues. Physical apposition of the wound edges is indicated by fine lines, pulling the edges into contact in the upper picture, but not altering tissue specificities (i.e., leaving the pieces unmoved, only bridging the range of states presently unpopulated by cells) in the lower image. New cells adopt tissue specificities intermediate between their neighbours', thus populating the fine lines and restoring the image, if not necessarily the physical geometry of the organism.

In experiments on the crayfish, Mittenthal shows that these "in-between-ness" connections *cannot* always be made *straight* by suitable rubber-sheet rearrangement of TSS, contrary to my conjecture of 1977. This may be a consequence of nothing more exotic than *curvature* of TSS: if TSS were like a steep hill on an otherwise flat plane, then between any two points astride the hill there would be *two* shortest paths (left and right along the hill's flanks) from one to the other through "in-between" cell states. Any representation that purports to straighten all such curves would thereby inadvertently superimpose these two, though they consist of distinct cell types. So a two-dimensional TSS cannot always be so defined that the linking curves appear straight. Nonetheless, straight-line connections in two-dimensional-space appear to suffice for the purposes of this Chapter, obviating the need to adopt a special "shortest arc" rule and the implicit metric to decide which path to take in a one-dimensional space.

We rewrite the rules now in three parts using this geometric language:

Rule I: Each little patch of cells is labeled with a state which is part of a smooth gradient of states across the tissue. This state is a point in a two-dimensional space of biochemical specificities topologically equivalent to a plane. Once established, this smooth map does not change. The two components of this "state" specification might be, e.g., the concentrations of two kinds of adhesive site on the outer cell membrane.

Rule 2: No exceptional point or state exists.

Rule 3: Cells proliferate at a rate determined by the separation of adjacent cells in TSS (the unlikeness of surface properties, for example). If they are initially quite different in tissue specificity (e.g., cells on the interface, connected by fine lines, then they begin mitosis. The tissue expands in physical space while its image in TSS remains unmoved but becomes denser with cells. Proliferation stops when the density of cells throughout the image has risen to a local norm, i.e., when the local gradient of tissue specificity has diminished to normal for each kind of tissue. There is no problem about endless proliferation at an unremovable phase singularity: no singularities occur. Just as in the original rule, new cells intercalate tissue specificities intermediate between their physical neighbours. (Note that we have yet to specify exactly what is "intermediate" between cell types that are not normally adjacent. This puzzle is pushed much closer to solution in Mittenthal's *Rule of Normal Neighbours*.

Applying the TSS Image Rules

No separate rule is needed for induction of a limb. The image in TSS of any ring of tissue is necessarily a ring, and by Rule 3 the tissue inside the physical ring must acquire (at least) the tissue specificities inside the ring's image. At an appropriate stage in development, cells will develop the structures by which their positions in TSS were named. If those should happen to include the distal parts of a limb (fingers, for example), then such structures (among others) arise when tissue specificities are interpreted biochemically, and we have a "limb." If the most distal structures are normally internal to more proximal structures in TSS, then amputation corresponds to ablation of a disk in TSS, leaving a ring. Blastema formation followed by wound healing corresponds to stitching across the empty disk fine lines along which new cells take up the missing tissue specificities, restoring the more distal organs. Replacing a left hand (L) by a right hand (R) on a left stump corresponds to cutting out the disk in TSS and replacing it with an identical disk. But remember that because of the inevitable L-R mismatch in physical space, tissue connections from the (R) central disk reach across to the opposite side of the (L) hole. We thus have a three-layered map across the distal core of TSS. The three layers are the existing hand and the cylinder of wound along the wrist that joins hand to forearm. The two new layers (the cylinder of wound proliferation)

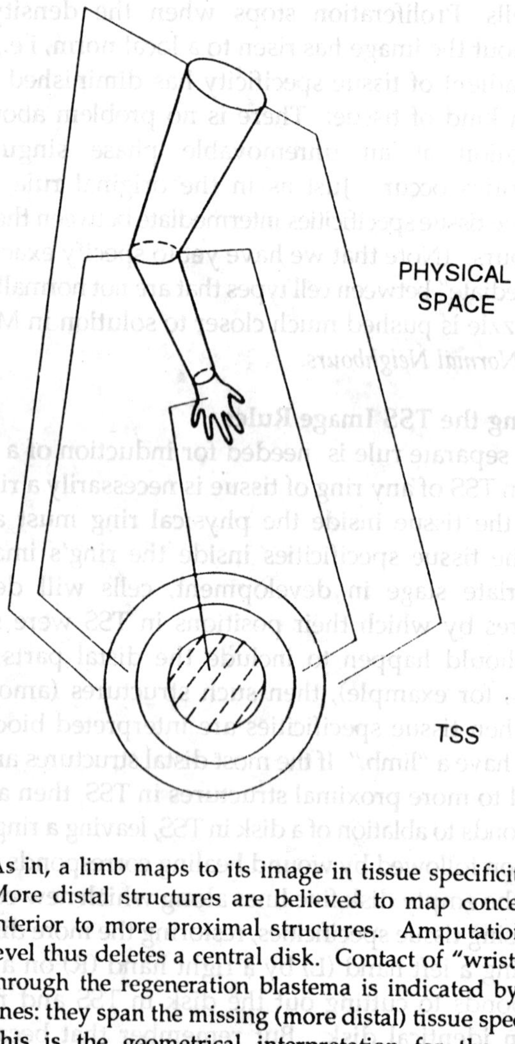

Fig. 3.11. As in, a limb maps to its image in tissue specificity space. More distal structures are believed to map concentrically interior to more proximal structures. Amputation at any level thus deletes a central disk. Contact of "wrist" tissues through the regeneration blastema is indicated by dashed lines: they span the missing (more distal) tissue specificities. This is the geometrical interpretation for the empirical generalization that mutilated tissues grow only more distal structures from the exposed edges.

have the same orientation, so that we get two replicate left hands. Or imagine that the arm is amputated at both ends. Both the proximal and distal blastemas must then span the middle ("hand") region of TSS. So mirror image hands must emerge (and do) at each end. This is the simple geometric essence of the rule of distal outgrowth. According to this rule, more distal structures regenerate from any stump even at the proximal end where, if regeneration were naively expected, a shoulder should grow.

This picture in TSS is the same (*not* inverted) for right or left limbs. Thus replacing a left hand (L) by a right hand (R) on a left stump corresponds to cutting out the (left-hand, L) disk in TSS and replacing it with another (right-hand, R) disk. The new disk looks exactly the same as the old in TSS: it covers all the same regions of the "inverse fate map," since it has all the same cell types in the same neighborly arrangement. But the reversal of handedness corresponds to a reversal in the correspondence between cell types and points in real physical space. This only becomes important at the boundary where the graft and host are trying to fit together: in real physical space, that ring of cell types in TSS corresponds to a clockwise ring of host tissue, but to an anticlockwise ring of graft tissue.

Because of this inevitable L-R mismatch in physical space, tissue connections from the (R) central disk must reach across to the opposite side of the (L) hole. Note that this is not quite crazy; to switch handedness, one must invert an organ through some axis, commonly the medio-lateral axis as in, but often (and with comparable effects) the antero-posterior axis or any intermediate. Each cell type at the junction finds itself adjacent to cells of the same antero-posteriorness, but opposite medio-lateral-ness. The new cells arising in the junction between graft and host must accordingly cover the range of medio-lateral positional values (tissue specificities) between opposite extremes. This range is spanned by the new cells on each side (and is of course already spanned within the graft itself). We thus have a three-layered map across the distal core of TSS. The three layers are the existing hand and the cylinder of wound along the wrist that joins hand to forearm. The two new layers (the cylinder of wound proliferation) have the same orientation, so that we get two replicate hands. Since that orientation is upside-down relative to the (R) graft, they are of opposite (L) type, the same as the host. The parity-conserving final result is a left limb with a left hand, but with a complementary L-R pair in addition. This strange, and (originally) unexpected experimental result derives from nothing more than an assumption of continuity(!).

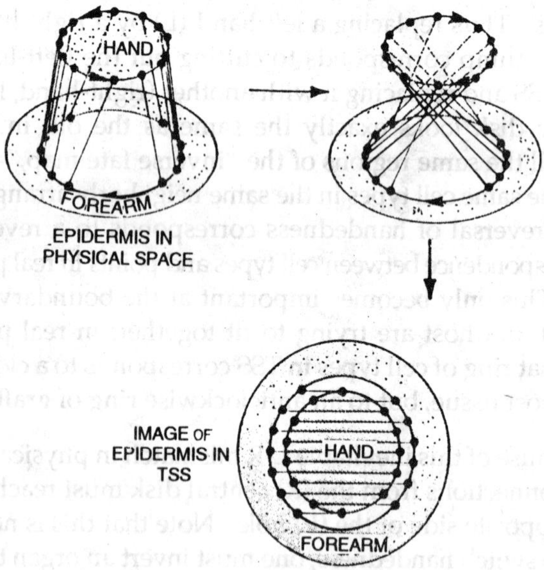

Fig. 3.12. The experiment is represented in the terms, rather than in terms of phase contours. In the upper left panel, a hand with clockwise polarity is inserted in a forearm with anticlockwise polarity. The separation of hand from forearm in a third dimension here is only for convenience of manipulation in the transitional second panel: the actual epidermis and its image in TSS are, of course, two-dimensional as in. In the lower right panel, the hand has been flipped to its standard orientation in tissue specificity space. Regenerating tissue, being suspended between mismatched left-handed boundaries, is now seen to span the "hand" region twice, inverted relative to the graft hand. Two supernumerary hands equivalent to the one removed are thus inevitably induced by the requirement of continuity alone.

Equivalent Metaphors and a Coat of Many Colors

Another way to articulate the TSS rules uses the language of *colour*. The diagrams in every paper about the clockface model would be much simplified, and both design and interpretation of experiments would be made transparent if only we could publish airbrushed colour diagrams as follows. Let the phaselike, circular aspect of tissue specificity be represented not in digits 1-12, nor in contour lines, but in *hues* taken from a colour wheel. On a colour wheel, red grades into orange grades into yellow grades into green grades into blue grades into indigo grades into violet grades into red grades into orange, and so on: without backtracking, one can progress smoothly and unidirectionally through a cycle of tissue states, as required. Proximo-distal-ness may then be represented by *saturation* of the hue, the most distal tissues being least saturated (gray) so that all hues nearby are compatible without discontinuity. Our labelling of tissue types then consists only of assigning to each tissue type (and so to each place on the organism) a colour: a hve and its saturation. Every organism, in this view, wears a coat of many colours. The colour is permanently imbued in each cell at mitosis, in cases of strict epimorphosis. Surgical interventions consist of abutting together colours that were previously separated by gradations of colour. If the colours now "run" and blend (as in morphallaxis), or if new fabric intercalates itself, adopting hybrid colouration compatible with its boundaries (as in epimorphosis), then we end up with arrangements of colour that satisfy all the phenomenology of regenerating and duplicating tissues, and, incidentally, the mathematical phenomenology of phases 1-12 (hues) and winding numbers thereof (complete colour wheels) and singularities (gray spots).

The phase circles of Glass may be traced through full cycles of hue, ignoring saturation. They can be constructed by drawing circles on the physical tissue and then following the circles' images (rings) in TSS. The number of singularities corresponds to the winding number of the image about an arbitrarily chosen origin for the angular co-ordinate system. In the coat-of-many-colours metaphor, the usual choice of origin would be gray, but this is an arbitrary convenience dependent on ambient lighting. Biologically, the usual choice of origin is the distal-most tissue. From my point of view, the arguments of French *et al.* and of Glass about circles, phase maps, and winding numbers amount to using a circle embossed on the organism as a means of book-keeping the folds and rotations of the regenerating tissue's image in TSS. Glass's calculational procedure is very useful here, because until one has a little practice, it seems awkward to visualize handedness in TSS maps. The essential feature common to all these metaphors is the mapping from a physically distributed tissue to an abstract state space of two dimensions with continuity properties like ordinary two-dimensional Euclidean space. The most general underpinning for such a mapping is a set of chemical substances arranged in the tissue along transverse concentration gradients. To take but one example, the local concentrations of any three distinct pigments define a three-dimensional state space. Any surface coloured by mixtures of those pigments is thereby mapped into that space. If the total amount of pigmentation (the sum of the three concentrations) is limited to a fixed quantity, then the surface is confined to the two-dimensional triangular plane of standard colorimetric charts. Colour mixtures happen much as in. If the "pigments" (or cell-surface antigens, etc.) should happen to mutually influence their own rates of synthesis and degradation then, as is well-known from the classical theory of reaction/diffusion structures, the chemical gradients may become self-stabilizing. The only experimentally demonstrable reaction/diffusion structure at present is the "rotor" in Belousov-Zhabotinsky reagent. Its structure, dynamics, and reaction to trauma may be expounded exactly in the manner of this chapter, its development into three dimensions resembles nothing so much as the embryology of simple organisms.

From the literature surrounding the clockface model (1), one gains the impression that proper understanding of the

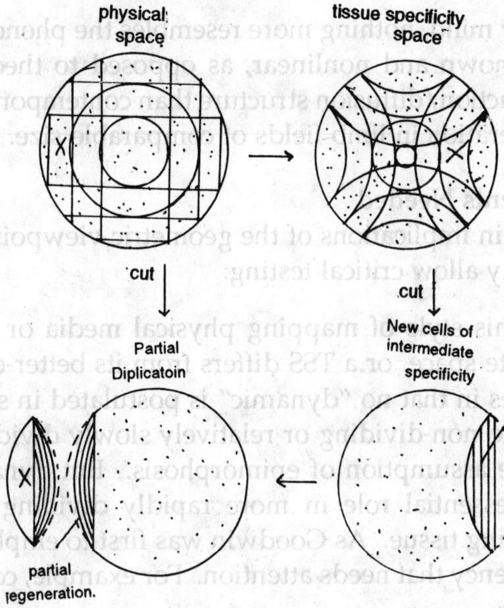

Fig. 3.13. (Legend: above = before; below = after, left = physical space; right = tissue specificity space.) The imaginal disk of a fly's appendage is painted with grid lines and with concentric contours of proximo-distalness, then mapped (right) into TSS. The "puckered" mapping is essential for what follows. Cutting off a chord in physical space (lower left) excommunicates that chord's lens-shaped image in TSS (lower right; the lens is drawn as though displaced from its initial tissue specificities only for graphic clarity). Supposing that each piece folds to close the cut edge on itself, new cells arise with intermediate specificities indicated by fine vertical lines: each edge grows back about half of the cell types in the severed chord. More severe "puckering" of the map would improve this partial regeneration and reduplication, but to guarantee completion we must cease to ignore the *dynamics* of specificity in TSS.

phenomena of regeneration somehow repudiates (in favour of the more modern preoccupation with cell surface chemistry) the classical and apparently sterile notion that chemical concentration gradients and reaction/diffusion structures underlie it all. Perhaps it is indeed time for a fresh perspective on these unanswered riddles, as old as the science of biology.

But to my mind, nothing more resembles the phenomenology of the (known and nonlinear, as opposed to theorized and linear) reaction/diffusion structure than contemporary reports on regeneration in limb fields of comparable size.

Experiments Needed

Certain implications of the geometric viewpoint adopted above may allow critical testing:

1. This style of mapping physical media or organisms into a state space or a TSS differs from its better-established prototypes in that no "dynamic" is postulated in state space. Applied to non-dividing or relatively slowly dividing tissue, this is the assumption of epimorphosis. But dynamics may play an essential role in more rapidly dividing young or regenerating tissue. As Goodwin was first to emphasize, this is a deficiency that needs attention. For example, consider the following:

(a.) Although French *et al.* address themselves exclusively to epimorphosis (old cells retaining their specificity), some amount of morphallaxis (respecification) may commonly occur in regenerating limbs. According to Maden's interpretation, this fact is assimilated by simply allowing our Rule 3 to govern wounded tissues even prior to cell division. This scarcely alters our diagrams. However experiments to elucidate the following two points are potentially more subversive.

(b) Every tissue starts as a tiny patch of indistinguishable cells which, as they grow, acquire divergent specificities. Developmental fields orginate, like the "big-bang" model of the expanding universe, from a singularity in TSS. At least at that early stage of development a "dynamic" *dominates* TSS. Kauffman *et al.* provide an alluring linear theory of this "dynamic."

(c) Observations on transected imaginal disks of the fruitfly argue for a strong tendency of tissue specificity to evolve more proximally. The qualitative observation

> is that both edges of the cuts produce the same new structures: one piece exactly duplicates and the complementary piece regenerates completely. The knife line though straight in the real world, is generally curved on the image of a disk in TSS, TSS being defined in such a way that inbetweenness follows straight lines so far as possible. No matter how the cut edge heals onto itself, the proliferating cells will adopt tissue specificities intermediate (parallel lines) between points on the knife cut's image in TSS. These intermediates cannot include the whole domain occupied by the excised piece if the disk's image is convex in this TSS— and it must be, or else intercalation between protrusive extremes of the image would produce tissue types outside the image, i.e., not normally present. So developmental fields are *convex* and cutting the edge off one leaves each part completely without access to some region of TSS occupied by the other part. Without a "dynamic," one piece must then fail to *fully* regenerate while the other fails to *fully* reduplicate. Transection experiments may thus provide the means to verify the existence of a "dynamic" and to quantitate the rates at which tissue specificity can change during normal growth and during regeneration.

Without a proximalizing "dynamic," there must be parts of any organism that *cannot* be regenerated: the adjacent most proximal parts of neighbouring developmental fields, located on the common rim of their convex images in TSS. By the same token, the kind of regeneration here considered cannot proceed beyond any symmetry plane: a bilaterally symmetric organism, for example, could regenerate at most half of its complete body. How do these limitations contrast with the facts of regeneration in radially symmetric organisms and in morphallactically flexible organisms such as *Planaria*?

Recent experiments using artifically constructed bilaterally symmetric limbs are particularly intriguing in this context. An unexpected progressive loss or gain occurs in circumferential positional values (hues, clock-phase values, TSS regions occupied). This may betray a slow underlying dynamic such as we inferred previously.

2. According to this coordinate-free representation, ablation of a feature on the symmetry plane of a bilaterally symmetrical organism (a tongue, a nose, a penis, a tail, the genital disk of a fly) corresponds to cutting a hole out of the fold of the two-layered image in TSS. Such a hole can heal in two quite distinct ways.

(a.) Closing horizontally (top-to-bottom contacts), the hole in TSS would be filled again. Thus we should anticipate complete regeneration.

(b) In contrast, a vertical closure (side-to-side contacts) would join mirror image tissues. Thus, no proliferation should ensue. Even if it did, the missing tissue specificities would not be recovered.

The results of diagonal closure must vary between the external results a and b according to the angle of the diagonal. This series of experiments may help to distinguish between geometric interpretations of tissues specificity.

For discussion of experiments recently undertaken along these lines, consult. One source of ambiguity at present is the lingering uncertainty about the actual (uncontrolled and unobserved) geometry of wound closure.

3. If one had to guess before observing, it might reasonably be supposed that an organism would respond to surgical challenges by intercalating positional values between given boundaries in the smoothest way. This surmise comes from the observation that paired phase singularities (left-right pairs of supernumerary appendages) are uncommon where uniformity of phase might alternatively prevail.

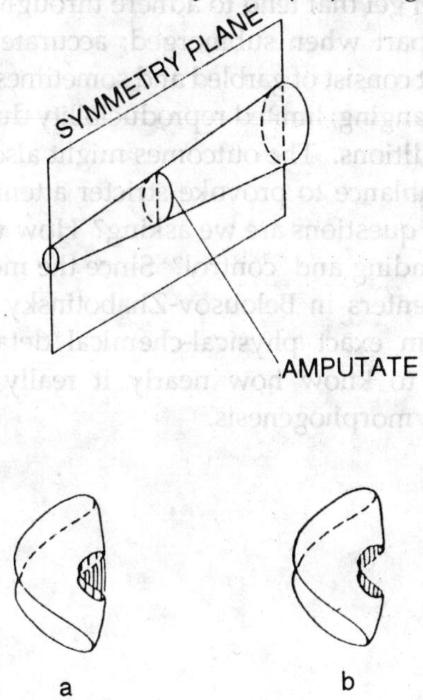

Fig. 3.14. As in, a symmetric appendage maps into TSS as a folded disk, its corresponding left and right parts *twice* covering each occupied specificity. Amputation deletes a central disk *straddling* the *fold line*. In contrast to the inevitable distal regeneration of asymmetric structures, this one may regenerate to any degree, depending on the exact manner of tissue appositions during wound closure. The extremes are shown in (a) and (b).

4. It seems not entirely proven by experiment that apposition of dissimilar tissues is required to elicit regeneration. It is possible that the external medium may under some circumstances serve as a "tissue type"? Where is its point in TSS, if so?

5. It would be of interest to treat a strictly non-living chemical "organizing center" as though it were, for example, a limb field, repeating many of the physical manipulations of classical embryology. Some of the practical difficulties are similar: Accurate separation and recombination of microscopic chunks of wet gel that tend to adhere through surface tension or to float apart when submerged; accurate observation of outcomes that consist of garbled and sometimes subtle patterns, constantly changing; limited reproducibility due to uncertainty of initial conditions. The outcomes might also bear sufficient formal resemblance to provoke stricter attention to strategic issues: What questions are we asking? How will the answers aid understanding and control? Since the morphogenesis of organizing centers in Belousov-Zhabotinsky reagent can be understood in exact physical-chemical detail, it might be illuminating to know how nearly it really does resemble cytochemical morphogenesis.

4

Morphallaxis and Regeneration

The term morphallaxis was applied originally in the study of animal regeneration, early in the 1900s. It defined a process where structure appeared to be regenerated by remodeling over a wide area within remaining tissues, so that complete pattern was restored without any localized zone of special growth being responsible for the replaced parts. Such regeneration can be seen in adult planarians and urochordates, but the precise roles played in it by special reserve cell populations, cell migration, and the alteration of differentiation and growth remain imperfectly understood. Recently, the term has been extended to initial development to refer to those regulative systems, seen in early embryos of many (perhaps most) types, where cells' contributions to the final pattern of the body are set by their *relative positions* within a cell sheet or a mesenchyme during the period across which determination occurs. The proportions and completeness of the pattern are thus independent of growth schedules and of the size of individual examples. This contrasts with patterns of cell differentiation built up by directed lineage, or by the local control of growth and cell fate by interaction among neighbours i.e., epimorphic processes. Some far-reaching mechanism of intercommunication or signalling within the embryonic tissue is obviously implied by this positional phenomenon, but although the idea that there is essentially just one form of such mechanism is forceful and attractive, we still lack evidence for this. Moreover there is, in evolutionary terms, less a priori reason to expect a universality of mechanism for control of multicellular patterns than for, say, the hereditary transmission of genetic information. In this sense, continued use of the term morphallaxis as if it referred to one as-yet-unelucidated mechanism may be disadvantageous.

The system is that which initially specifies the body plan in the early vertebrate (specifically, amphibian) embryo. Our own membership of the zoologically very homogenous vertebrate group gives an incentive to attempt to understand the early development even when we currently lack some of the technical advantages evident for other groups, notably insects. Certain vertebrate systems have a classic status as experimental objects, probably because of ease of access for surgical manipulation. This has resulted in an extensive backlog of geometrical knowledge about the fate map of material and the cellular movements in undisturbed development, such knowledge being prerequisite for any approach to analytical rigor in experiments of the type to be described.

THE MEDIO-LATERAL DIMENSION

The Anatomy of the Developing Pattern

The amphibian mesoderm develops by immigration of cells from a surface or sub-surface equatorial zone, the marginal zone, of the blastula. The future dorsal midline cells lead the process but the zone from which tissue is recruited soon comes to be annular, so that a bilaterally sysmmetrical, cylindrical cell sheet develops with an extended dorsal side. Keller Nakatsuji and co-workers have studied in fine detail the origin, fates, and behaviour of the mesoderm cells during normal gastrulation and pattern formation. By the time the medio-lateral pattern of its future contributions to the body is being established, it is one or a few cells thick, and situated between the neurectodermal layer and the yolky endodermal mass. Cells are continually added to form the cross sections of more posterior levels of the cylinder (and thus body axis) from an annular zone around the closing blastopore.

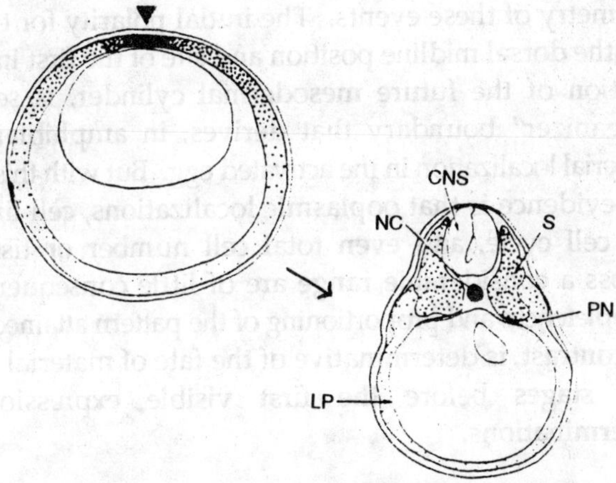

Fig. 4.1. Schematic cross-sections of midgastrular and early stages of amphibian development. The middle, mesodermal cell layer and its four derivatives in the developed meiolateral pattern are stippled with decreasing density from middorsal (arrowhead) to ventral. Stippling density indicates differentiation tendencies at the early stage where the mesoderm is essentially a monolayered cylinder; see the later territories notochord (NC), somite (S), pronephros (PN), and lateral plate (LP), after the dorsal convergence movements and induction of the overlying nervous system (CNS). Except for dorsal midline positions, however, determination does not occur until relatively late stages (i.e., neurula and early tailbud).

The four territories comprising the archetypical vertebrate medio-lateral pattern are the mid-dorsal notochord, the paired somite-forming columns, paired pronephric zones and the lateral plate mesoderm with blood-forming islands ventrally. The tissue to be devoted to these structures is established as zones, of particular relative extents, within the cross-section of the mesodermal cylinder just described. The total field between the extremes is in the order of 100 cells and 1.5 mm within the cell sheet. Thus with four stripes and bilateral symmetry as trivial added features, this is an instance of the problem of the regulating *"French flag"* used by Wolpert in his general treatment of the concept of pattern formation. The geometry of these events. The initial polarity for the pattern, i.e., the dorsal midline position and site of the first immigrating portion of the future mesodermal cylinder, is set from an "organizer" boundary that derives, in amphibians, from a material localization in the activated egg. But with this exception, the evidence is that ooplasmic localizations, cell lineage, and the cell cycle, and even total cell number or tissue extent across a considerable range are of little consequence for the completeness and proportioning of the pattern attained. Position, by contrast, is determinative of the fate of material until quite late stages before the first visible expression of the determinations.

OUTLINE OF CLASSICAL FINDINGS AND CONCEPTS

Cell Fate Determined by Position

The term *determination* as used of embryonic tissue must always be defined operationally; that is, in terms of what has been done to the cell to test their state of commitment, or restriction in developmental potential. Unclearness about the implications of particular observations, in assessing the state of determination of parts of the embryo, has dogged vertebrate developmental biology ever since the period of its classic works. Thus one surprisingly often encounters views to the effect that the body plan of amphibian embryos is built up because the successive cleavage products, throughout the egg, inherit different subprogramms in the form of "ooplasmic" precursor localizations, or because small groups of early cleavage products are set aside (e.g., by binary sequential decisions) as the only possible founders of parts of the pattern, as is seen in insect embryogenesis. It is appropriate here to summarize the evidence for three assertions, namely (a) that cell potential for most parts of the mesodermal body pattern is liable until late stages of gastrulation, (b) that the property of acting as a pattern-controlling boundary "the organizer" is restricted to a particular group of cells that normally develop the mid-dorsal territory, and (c) that cell-cycle related changes of differentiated state, and/or differential migration and adhesion among determined cells cannot make more than local and late-acting contributions to primary pattern establishment.

Despite Driesch's original characterization of the unknown mechanism of morphallactic development as a "harmonious equipotential system" there are few embryo types which long remain in the condition where all large coherent parts, when isolated, will adjust the fates of their cells to form complete, normally proportioned editions of the body plan. Vertebrate embryos are indeed among the most highly regulative in this way, but from the onset of cleavage in amphibians, isolated pieces (e.g., halves) will only regulate and form substantially normal pattern if they include material from the restricted region, the organizer, whose fate is to contribute to dorsal mesoderm and head endoderm. As already mentioned, properties that are the precursors for this region are first localized within the egg near the onset of development. If an embryo fragment has the organizer region near one of the edges that must reheal to restore its geometric continuity (i.e., if a presumptive lateral, rather than a dorsal, half embryo is created), the achievement of normal pattern is unreliable from all stages and seldom found from operations as late as late blastula (5000 cell) stages. But this loss of regulative capacity does not mean that an advanced degree of determination has taken place among the individual cells of the embryo. Large mesodermal isolates cultured from much older (50,000 cell) gastrulae often develop according to presumptive fate only (i.e., no regulation), but even this does not imply determination among their cells at the time of explanation. Such observations tell us only that intercellular communication systems that normally regulate pattern (including any biomechanical as well as biochemical factors) have dynamics such that a spatially complete set of events can no longer be achieved from the starting conditions obtaining within the isolate. No cellular "decisions" need have occurred, and such isolates, placed as part of another embryo that retains the appropriate dynamics (effectively, one that possesses the organizer), may deviate from their original fate in whatever direction is required for the production of normal pattern in the host.

Much about the communication system that regulates pattern might still be learned from a systematic exploration of the times at which each sector of the marginal zone other than the organizer loses the capacity to be assimilated into the normal pattern, when relocated to particular sites. Spemann and co-workers were responsible for a body of work which led to concepts, about the dynamics of such development, that have essentially been inherited by a majority of vertebrate embryology laboratories today. Their empirical discovery of the organizer and of the induction of the nervous system were undoubtedly achievements of crucial importance, but their theoretical legacy may have done much to retard the subsequent progress of understanding.

It was found that one specific cell group, when relocated to remote, presumptively ventral parts of a host embryo, was able to cause development of a second edition of the body pattern, centered on itself but sharing the host material with the pattern of original orientation. This group included the dorsal midline of the zone of mesoderm immigration (dorsal lip) and future head endoderm at early gastrula stages. The graft developed to make its normal, dorsal, and anterior presumptive contribution in the second pattern, while the extent of host participation was variable but could, under some circumstances, approach a 50% subdivision of the available material with the original pattern. To confine ourselves momentarily to the mesoderm, the real subject of this chapter, it can be deduced from Spemann's writings that he believed that the respecification of the fates of material, involved in this phenomenon as well as in that of regulation to wholeness by early dorsal or lateral half embryos (see above), was essentially on a geographical or relative positional basis. Any special stimulation of cell division and thus new tissue production played at most a minor role. Contemporary work has shown

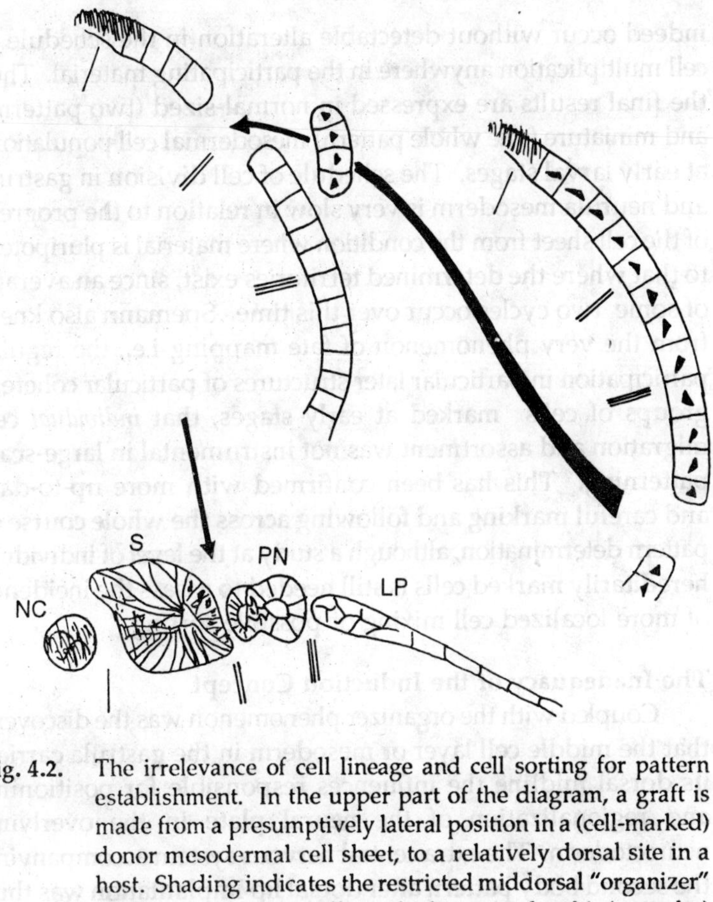

Fig. 4.2. The irrelevance of cell lineage and cell sorting for pattern establishment. In the upper part of the diagram, a graft is made from a presumptively lateral position in a (cell-marked) donor mesodermal cell sheet, to a relatively dorsal site in a host. Shading indicates the restricted middorsal "organizer" regions, the only ones effectively determined at this (gastrular) stage. Some 24 hours and two cell cycles later, the formed pattern is observed in the host (lower part of diagram), showing descendants of the implanted cells to have remained as a largely coherent patch with few outliers, and to have been determined in harmony with their new positions in the cell sheet as a whole. Index lines mark the presumptive and actual frontiers between pattern territories. Without extensive cell sorting, the pattern could not be built up on a lineage basis within the number of cell generations available.

that these two major regulative adjustments of pattern, to the mirror-symmetric double dorsal and to the small and whole, indeed occur without detectable alteration in the schedule of cell multiplication anywhere in the participating material. Thus the final results are expressed in normal-sized (two patterns) and miniature (one whole pattern) mesodermal cell populations at early larval stages. The schedule of cell division in gastrula and neurula mesoderm is very slow in relation to the progress of the cell sheet from the condition where material is pluripotent to that where the determined territories exist, since an average of some two cycles occur over this time. Spemann also knew from the very phenomenon of fate mapping i.e., the regular participation in particular later structures of particular coherent groups of cells marked at early stages, that *individual* cell migration and assortment was not instrumental in large-scale patterning. This has been confirmed with more up-to-date and careful marking and following across the whole course of pattern determination, although a study at the level of individual hereditarily marked cells is still needed to assess the incidence of more localized cell mixing or possible sorting.

The Inadequacy of the Induction Concept

Coupled with the organizer phenomenon was the discovery that the middle cell layer or mesoderm in the gastrula carried its dorsal midline the influences responsible for positioning and regionalization of the neural plate in the overlying neurectoderm. The extra central nervous system accompanying the second body pattern after dorsal lip implantation was thus a natural result of incorporation of an organizer into the mesoderm beneath it, being a territory *induced* within the neurectoderm. As implied by Wolpert, it would have been well for developmental biology if the term, *induction,* had then been restricted to events like this evocation of the nervous system from neurectoderm by contact with mesoderm; that is, to the transfer of information determining a pathway of development between one cell layer and another. As it was, Spemann and his school expanded the concept to refer indiscriminately to all hypothetical signals, emanating from particular determined territories of pattern to evoke other determinations in neighbouring material. The spatial organization of development came to be seen just as a sequential hierarchy of inductions, with failure to distinguish conceptually between processes believed to occur along the plane of one cell layer or tissue and those occurring between two opposed layers or tissues and co-ordinating the patterns of differentiation between them. This has remained the conceptual framework for most embryologists. Thus most textbook accounts of the grafted amphibian organizer phenomenon fail to distinguish between the direct effect of the graft itself, which is to set a new boundary for the dynamic system organizing the pattern of mesodermal territories, and the accompanying induction of a second central nervous system with this new dorsal midline of pattern as a secondary expression.

We are here concerned only with events controlling primary pattern, which arise in the mesodermal layer. It can be seen intuitively that without some ancillary concept corresponding to negative feedback or self-limitation, models based on hierarchies of induction fail fundamentally to explain the salient features of morphallactic regulation, the size-independence of the completeness and proportions of pattern. Briefly, if cells on finding themselves to be developing as type A manufacture a specific substance which diffuses or is otherwise exported into the surroundings, and causes there the development of cell type B, and so on, there is no way that the system as a whole can take account of the overall space available in setting the spatial scale of the territories devoted to A, B etc. The kinetics of manufacture, diffusion/export and breakdown of signals and of cells responses must be part of the species metabolic equipment, and not naturally susceptible to feedback modification against individual embryo size without a special mechanism. It is furthermore hard to see what such a feedback mechanism might be. Without it, the intrinsic property of sequential induction is to use up a particular amount of tissue or space for each particular territory induced, until either pattern is completed and the remaining space makes up the final territory, or the process runs out of tissue and pattern remains incomplete. Spemann, and most experimental embryologists writing since, while aware that vertebrate embryos avoid such incompleteness or variable proportioning over a wide, experimentally induced range of sizes at pattern formation, have been unable to confront this fact analytically.

The Testable Hypothesis that Positional Information Regulates the Pattern

Several biologists working with invertebrates, earlier this century, had tried to encompass morphallactic regulation in development and regeneration within a clear theoretical framework. In recent years, Wolpert has proposed a clear and, in principle, testable concept in the hypothesis that *positional information* underlies the control of pattern. He further proposed that this principal is universally used, and even that a universal mechanism underlies it in the development of regulative embryos across the animal kingdom. This vigorous and challenging notion has caused a wealth of new work, and undoubtedly stimulated the re-awakening of interest in the fundamental problem of pattern. It is currently in danger of degeneration or "dilution" however in that wherever pattern develops, then cells are said to have interpreted "positional information" and the two phenomenon are treated as almost synonymous. There is, accordingly, a certain impatience with the concept among experimental scientists. This is unfair, since although cast in abstract terms, the real scientific value of the hypothesis lies in its precise postulates about the form of the mechanisms, and thus its predictions which are potentially refutable in an appropriate test. It is not a *necessary* concept, since patterns could form and be regulated in other ways.

The abstract hypothesis is that normal pattern is achieved, despite any earlier disturbances which relocate cells within the field or create abnormally sized fields, because a communication system operates through the entire tissue. This system ensures that some cellular variable, whose local values set cells states to form the pattern territories, is distributed as a gradient with absolute boundary levels at the extremes of pattern and a particular, monotonic profile in the space between. The variable concerned might be a substance concentration or ratio of such concentrations or a range of cell surface states or configurations. Boundary states are set up by early localizations in the egg structure giving special properties to regions, or by symmetry breaking processes, and the consequent boundaries are the organizer regions seen in experimental embryology. Correct pattern depends on the appropriate interpretation of the gradient values by cells in terms of epigenetic action. There is then a family of more or less plausible hypothetical, but concrete physico-chemical systems whereby such a positional gradient might be set up and regulated, bearing in mind that normal pattern depends upon normal boundary values for the signal state and a normal gradient profile between them.

The application of the formal idea to the pattern under discussion, and then the principal model gradient mechanisms that have been proposed. These are diffusion from a focal source with degradation elsewhere, and reaction/diffusion models. The latter propose that autocatalytic processes of short-range diffusion produce a particular state of cellular activation as a local peak near one boundary, but lead to a larger-range diffusion gradient of an inhibitor of their own activity. The profile of either activation or inhibitor level may be used as the positional gradient. Diffusional transfer of substances of appropriate molecular weight to carry certain biological specificity might be consistent with the formation of gradients on the space and time scales (10 hr — 1 mm) that are required. The possibility of a series of cell surface states modified by neighbour-interaction, as a positional signal, exists but has not been formally modeled due to our profound ignorance of the machinery involved. The amphibian mesoderm system offers no evidence from grafting experiments for more than one organizing boundary, the organizer of Spemann. Ventral tissue experts no reciprocal effect to change the fate of its surroundings if grafted, but is assimilated into normal pattern if given time.

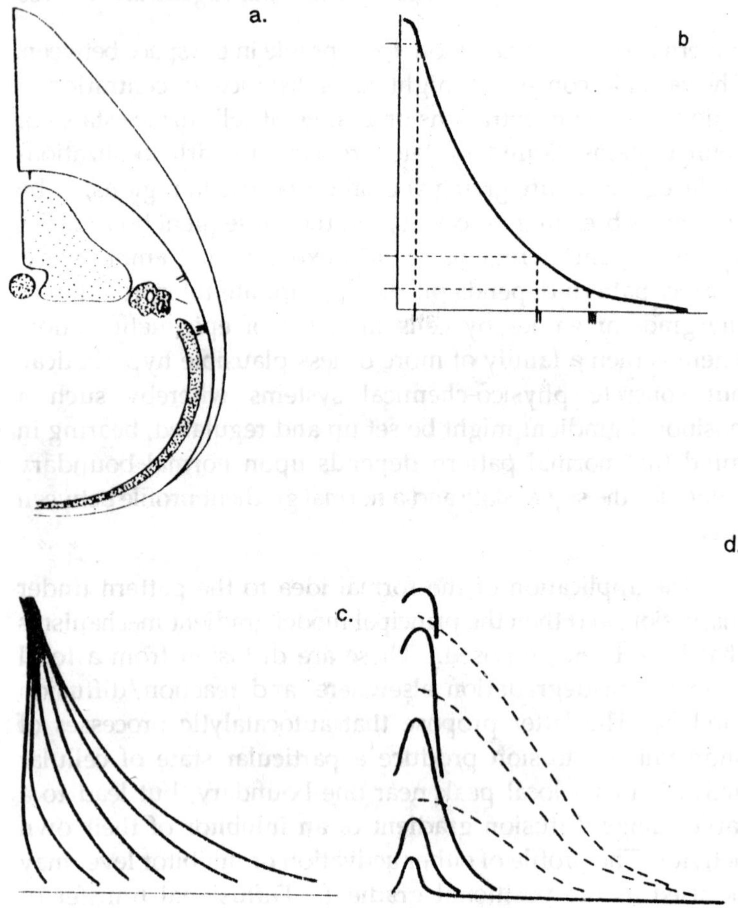

Fig. 4.3. The idea of positional information in media-lateral pattern formation. (a) The formed mesodermal pattern *in situ* with a gradient in positional signal extending around the cell layer from dorsal to ventral. (b) The formal idea of the gradient extending through the earlier cell sheet, showing threshold levels that occur at positions that become the frontiers between territories, devoted to the pattern parts. The proportions of these territories, in terms of cell number, depend upon the completeness and shape of the gradient. (c) A source of fixed concentration of a morphogen at one

boundary, and five timepoints on the way to equilibrium for a gradient maintained by diffusion/destruction elsewhere in tissue. Thus, if the system is not growing around the time of pattern formation, such a gradient only exerts good control over proportions when utilized near its (late) equilibrium state. (d) a reaction/diffusion created positional gradient. The solid profiles represent five timepoints in the slow spread of an autocatalytic activated state of metabolism, first arising in tissue near one boundary. Between the second and third timepoints, the activation exceeds a threshold leading to the local production of a more diffusible morphogen which spreads into surrounding tissue to form a gradient (dashed profiles). This morphogen preserves its own gradient distribution by acting as an inhibitor of initiation or auto-catalysis of the activated state in cells which have not yet passed an activation threshold, and thus limiting the spread of the activated region which alone produces it. At equilibrium, either the activation or the inhibitor profile could be used as positional information.

One area of test for the hypothesis concerns limits on the performance of these model physicochemical systems in producing complete and normal gradient profiles (thus complete patterns) under various abnormal conditions such as reduced size and bipolarity. Alteration of reaction kinetics and diffusibilities of the constituent molecules in any particular such system, in relation to individual embryo size, is impermissible because implausible; regulating behaviour must come, if at all, from the system dynamics. Does the performance of embryos under various circumstances match what plausible models can be shown to produce? The exploration of possible performances by model gradients from computer simulation or analysis is hardly exhaustive enough for quantitative observations about behaviour of real patterns to be critical tests of the validity of positional information. A much stronger test concerns the postulate, distinctive to this theory, that there are no specific interactions between emerging pattern territories, in the control of their relative proportions. Such control is stated to occur automatically from the independent positioning of the divides between the territories according to positional signal levels in the gradient. This suggests quite critical experiments in that specific, determined territories that do not include the organizer can be grafted into the cell sheet in which pattern is to be determined in somewhat younger embryos, and the effect upon sizes of the territories subsequently formed in those embryos can be measured.

The Pattern in Miniature Embryos and in Those with Two Organizers

The region of the long axis, in tailbud stage larvae, where the proportions and overall scale of the pattern can be assayed as soon as histodifferentiation has revealed the character of the component cells. Overall proportions of the total cells assigned to each of the four territories are estimated by counting the nuclei seen in each structure in a regular subset of the series of precisely transverse histological sections cut at standard thickness. Meaningful comparative results are obtained without correcting to absolute cell numbers present in each tissue, because pilot investigations reveal that nuclear size and internuclear distance in the various tissues, in the principal planes of section, do not differ as between normal and experimentally small embryos from within an egg-batch (i.e., sibling embryos). That is, their tissues differ in overall scale (cell number), but no significantly in texture. The rate of cell cycle between the period of determination of pattern and that of pattern assay is either zero or slow in the various territories and again there is no evidence for differences as between normal and experimental embryos during this interval, which lasts some 24 hrs. or 72 hrs. from the middle of the period of actual determination, in the fast-developing *Xenopus* and the slow-developing *Ambystoma*, respectively. The sampling involves several thousand cells per embryo, and repeat sampling of individual embryos gives very comparable results in terms of percentage of the sample population found in each structure. The control data have revealed that as expected, normal embryos, especially sibling sets, achieve rather constant proportions for the pattern. More remarkably, proportions are almost indentical in the two species used in the study, which represent the two major amphibian subclasses, and which differ about two-fold in the mean numbers of cells between medial and lateral extremes of pattern in cross-section (*Xenopus laevis*, the clawed frog, c. 70 cells, *Ambystoma mexicanum*, the axolotl c. 140 cells). The assessment of performance of the regulative system in this work is based on comparisons between synchronous, sibling sham-operated and experimental embryos in matched sets.

One completed study has assessed the response of the pattern specification system to the challenge created by two different abnormal circumstances. In the first test, presumptively dorsal part-embryos were made by removing ventral blastomeres (early cleavage products) from young blastulae and causing them to round up and heal. Morphologically small gastrulae and neurulae are produced which develop to tailbud larvae in which the mesoderm contains as little as 25% of the normal cell number, thus some 40% of normal cell numbers along the medio-lateral dimension. In the second test, late blastula stage embryos were provided with a second organizer (see later section), midventrally, by grafting a dorsal blastoporal lip from an early gastrula into a site made by removing a few cells from the ventral marginal zone. The new organizer diverts the development of some third or more of the total mesodermal cylinder into a new axial pattern having full cellular continuity with the "host" pattern at their adjoined lateral margins. Thus a double dorsal, mirror-image duplicated pattern is produced, involving a normal-sized cell population in total cross-section. These operations and their results are described elsewhere in full.

In single pattern of reduced dimension down to 40% of normal cell number, regulation achieves proportions for the four territories that are statistically indistinguishable from normal. The normal precursor material for lateral territories (even including pronephros) has probably been entirely removed in the more extreme examples. "Over-regulation" appears to have occurred. In patterns of the second experimental type,

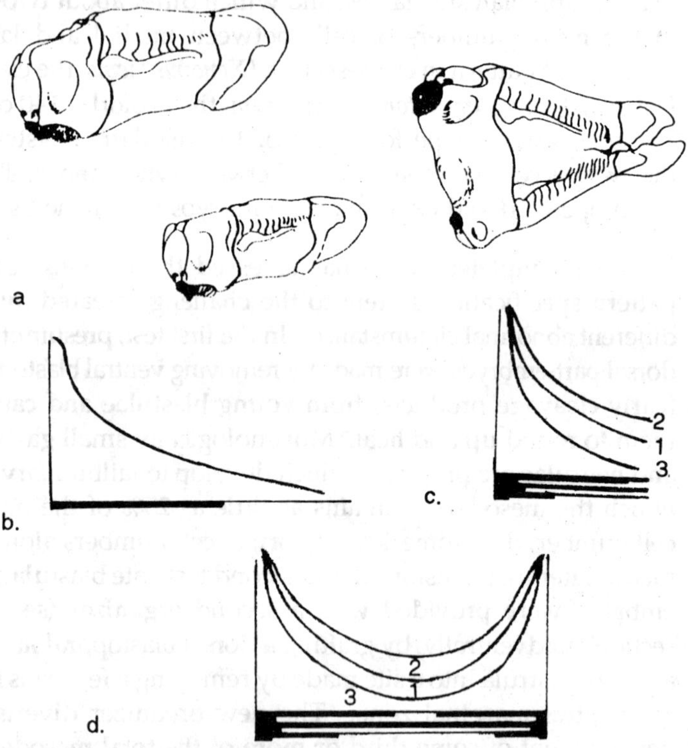

Fig. 4.4. Quantitive assessment of regulative performance by real embryos. (a) Normal, experimentally small and double dorsal organizer grafted embryos seen in fronto-lateral view. Heavy lines mark cross-sectional positions between which the standard estimates of pattern proportions by cell counting are made. Somite boundaries and pronephros can be seen in outline beneath the skin of the assayed area. The head contains a small proportion of the mesodermal cells in a complex set of structures, while the tailbud is the site of delayed differentiation and the first true growth, (b) An arbitrary morphogen gradient profile, taken to specify the normal pattern in a normal-sized mesoderm (the four thicknesses of the baseline), by a particular physico-chemical system. Performance of regulation in small (c) and double-dorsal (d) embryos. Profile and baseline versions one and two in these diagrams show the performance

to be expected of almost all models for diffusionally maintained morphogen gradients, on the positional information hypothesis of strict gradient interpretation. Profile and baselines versions three on the other hand, depict the regulative performance actually observed, and the profile of the gradient that would be implied, on a positional information mechanism for pattern specification. The essential discrepancy is that, empirically, there is little or no loss from territories far from the "gradient source," even though there is no evidence for control of the pattern from more than this one organizing boundary.

---

the tissue lines between two organizing boundaries where notochords have differentiated, and normal cell numbers are involved. Regulation produces mesoderms where the total percentages devoted to the four territories, across the two opposed patterns, lie much closer to control values than might be expected. Since the presumptively lateral third of the cell sheet has been reorganized in relation to a new dorsal midline, loss of lateral representations and the joining of two incomplete pattern might be expected. This is largely avoided, although lateral plate is statistically under-represented. Most strikingly and unexpectedly, the two patterns *both* contribute to this avoidance of lateral loss in the territory between them, since both show smaller cell populations than normal devoted to pronephros, somite, and notochord. The new organizer, mid-ventral in the original blastula/gastrula, has caused a scaling down of the positions of boundaries between territories in the original, host-controlled pattern. The phenomenon occurs most with respect to pronephros and somite territories, but to a significant extent even in "host" notochords, whose territory we believe to become fixed soonest after the onset of gastrulation.

These relationships and their implications for the positional information theory are represented. The curve represents the profile of a signal gradient whose levels specify normal pattern proportions at normal tissue size (shown along the baseline as cell numbers) on the positional information hypothesis for this system. A profile from a concave family is selected simply because plausible model gradient mechanisms characteristically produce such curves, but any monotonic form would do as well. The proportions for the pattern are known at some 4%, 40%, 10%, and 46% for notochord, somite, pronephros, and lateral plate/blood island respectively, with 10-20% coefficient of *relative* variation in each. We can thus mark, on the ordinate, threshold signal levels which would establish boundaries between territories so as to divide up the tissue space or cell number (baseline) in the appropriate ratios. The response empirically seen in small and in doubly organized mesoderms is shown as version 3 in terms of cell number along the baseline. If true regulation of an interpreted gradient were to underlie this response, therefore, the gradient profiles achieved would bear the relationships to the normal one that are shown in curves, and because of the new positions of boundaries. No gradient models having one organizer, as a signal source, with diffusion and degradation of signal elsewhere, can replicate such changes of profile in response to simulated equivalents of the biological experiments described. Few plausible models based on the more dynamic reaction/diffusion mechanisms (see later section) achieve the required behaviour either. Profiles are produced which more resemble numbers 1 and 2 of the figure, which when interpreted would produce baseline pattern versions 1 and 2, corresponding to loss of more lateral regions of the pattern, and even an associated expansion of medical territories.

A particularly challenging feature of the present biological pattern is that a restricted territory, the pronephros, far from either of its borders, is both rescaled and appropriately positioned in small versions. A steeply non-linear curve corresponding effectively to a peak of signal level near one border, the most usual reaction diffusion component that actually responds to change of field size by repositioning the peak shoulder, would not be adequate here. The need to produce the implied regulation with respect to a profile that would show slope, thus carry positional information, across most of the field, imposes even greater stability and plausibility problems on such models. In fact, only very narrow ranges of parameters for the simulated systems result in anything like effective behaviour. Certain *"solutions"* to the problem of steepening gradients by feedback from reduced overall scale of the system have been suggested. Thus, it is proposed that by various ancillary signals which are functions of tissue size, the diffusion terms or those describing the kinetics between autocatalytic and inhibitory components of the morphogen system are appropriately adjusted to restore the necessary gradient at equilibrium. All such "solutions" could be designed to work, by a human engineer. For the biologist, however, the intellectual choice would seem to lie between continuing to use such models, meanwhile wondering about their plausibility for quite basic and primitive instances of biological regulation, and on the other hand, turning to other classes of model which suggest new experiments. A more detailed discussion of the implications of the results just described, for various positional information models, appears elsewhere.

The Hypothesis that Serial Diversion of a Wavefront of Determination Regulates the Pattern

Biological patterns like the one we are considering have two main features, and an appropriate model must explain both of them. The territories that are determined are correctly ordered in space, and follow the *"rule of continuity"* when one edition of the pattern joins another in tissue, in that territory B is seldom omitted to leave A and C adjoining. Pronephros is occasionally absent in small amphibian patterns, but this may be because its visible expression is subject to a threshold number of determined cells being created. In addition, the relative numbers of cells assigned to the territories are quite highly regulated. There follows the outline of a model which departs from the positional information idea but which would explain the data thus far. It is related to previous ideas of Rose and of Meinhardt and also marks a reversion towards the original conception by Child of physiological gradients and dominance hierarchies controlling development.

The model involves a wavefront of cell determination passing across the available tissue, organized by a primitive gradient or by propagated cellular "activation" from the organizer, and a set of specific signals which divert determination from one course towards another according to a logical pattern. The organizer is assumed to control only an overall polarity within the embryo's tissue, this polarity being expressed as graded rates of physiological progress towards determination by cells, smoothly ranked from medial (dorsal) to presumptive lateral (ventral) extremes. Such control could be via a diffusion gradient of a signal state, declining smoothly with distance from source and setting local developmental rates, or else via early propagation of an activated cellular state, in the form of a wave that starts or accelerates progress towards determination, spreading from its origin at the autonomous organizer region. This spatial communication system need have no further, accurate informational function such as would be required of regulated positional information. It could therefore be a primitive and plausible diffusion-controlled or propagated signal, lacking regulative capacity against scale variation and maintaining only the direction and continuity of the wavelike sequence of development irrespective of precise position of material in relation to the boundaries of the pattern to be formed. If of gradient form, it could assume profiles 1 and 2 under the conditions corresponding while if of propagated wavelike form, its transit time would be proportional to tissue distance traversed rather than of particular absolute duration in both normal and small embryos (see later section).

Normal ordering and relative extents of the determined territories are assumed to follow because of the logic governing access to the various determinations by the cells. An early biosynthetic product of cells which have advanced into each determined state is assumed to diffuse rather rapidly into the rest of the tissue and to prevent, at a threshold concentration, development towards that same state by *"younger"* i.e., slower-developing or as-yet-unactivated cells that receive it. Thus in normal circumstances all cells that can develop fast enough move initially towards notochord determination. But a diffusable signal, produced specifically by cells that have become committed to the notochord state, builds up in the system as a whole until it attains a level that diverts less advanced cells towards the next state, somite. Such a process might be termed the *serial diversion* of a wavefront of *determination,* to produce pattern. Each specific signal is assumed to be so diffusible or transportable within the cell sheet as a whole that its level or concentration is a negative function of the size of the mesoderm as well as a positive function of the numbers of cells producing it. It thus acts as a sensor of the proportion of the total tissue that consists of cells which have been switched to produce it. The logic of proportion control is somewhat similar to that of the regulation of a single frontier position by a reaction-diffusion system, and the whole mechanism is like a linked sequence of such events that avoids the regulatory problems associated with positioning all the frontiers by one informational gradient. Due to more rapid building up each specific signal to threshold, the wavefront of determination will be diverted in character more frequently, marking off smaller territories between frontiers, when the cell sheet is smaller than normal or is being simultaneously invaded by another wavefront originating elsewhere as in organizer grafting. If the organizer is in fact the source for a primitive gradient, rather than the origin for propagated activation then the gradient is assumed to produce a wavefront of development simply by having smoothly graded, at some prior time, the rates of maturation in cells across the embryo. The conceptual separateness of the wavefront of determination (spatially organized), and then the succession of territory-specific signals that monitor the proportions of each pattern part is crucial to the understanding of this hypothesis.

The positional information hypothesis is distinguished from the foregoing one by its strong proposition that there are no specific interactions between territories during pattern control.

It has been reported that in double dorsal patterns, even the two notochords and pairs of somite columns, far apart in cellular terms, can affect one anothers' sizes. If specific feedback signals that exert effects on determination are to account for this, they must be sufficiently mobile so that significant spread throughout the mesoderm occurs within fractions of the time-course of pattern determination. This has consequences that suggest experimental studies. Firstly, a quantitative study of the time course of dissemination through cell sheets of labelled molecules of various characters and size is long overdue. Such a study could be carried out by following dissemination with time from small grafts of tissue heavily labelled with marked molecules, that had been injected into a blastomere of the original donor embryo. We have currently no idea how this time course would compare with that for free diffusion in aqueous space of the same dimensions, because the effects of restricted communication between cells via gap junctions, and possible effects of active transport or mixing in intracellular space, could modify effective "diffusion" coefficients in opposite ways. Figures derived from such studies could offer plausibility estimates for a variety of models, but no more than this. The biosynthetic arrangements underlying actual signals used in pattern regulation could give their progress through tissue a character intermediate between that of diffusional spread (proportional to distance squared) and that of a propagated state or wave (proportional to distance).

Specific Suppression of Territories by Heterochronic Grafts; a Test of the Serial Diversion Hypothesis

Another study suggested by the serial diversion model involves transplantation of large pieces of already-determined territories, from older embryos, to ectopic locations in gastrulae whose own patterns have yet to be determined. With certain donor-host age relationships, such grafts should exert specific suppressant effects upon the percentage of the host mesoderm finally devoted to the homologous territory. Any significant observation of such a phenomenon would constitute strong evidence against positional information as the effective control mechanism for the pattern. Experiments have been designed to search for such effects and the results of a study in progress are outlined in this section.

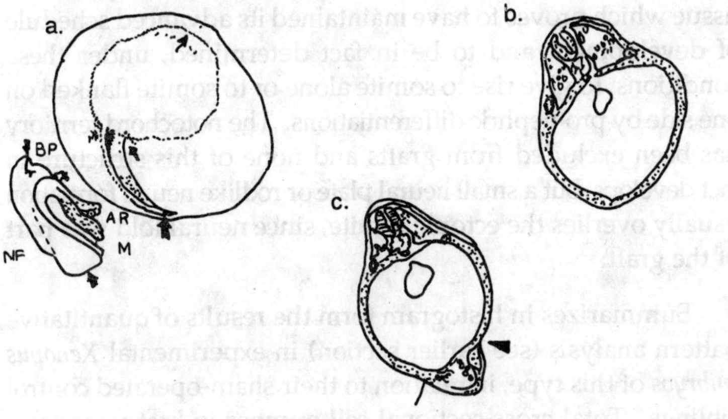

Fig. 4.5. The heterochronic grafting operation. (a) The operation whereby a dorso-lateral sector of neurula, including neural fold (NF), presumptive somite mesoderm (M), and the archenteron roof (AR), joined at the blastoporal zone of mesodermal recruitment (BP), is grafted to a ventral site made by cutting a slot in the ectoderm back to the marginal zone in a mid/late gastrula host. Light arrow marks dorsal middle of the host (which is seen from the yolk-plug aspect), while heavy arrows show how cell layers are matched up; (b) control sham-operated, and (c) heterochronically grafted tailbud stage embryos in schematic transverse section, to show how pattern parts are histologically recognizable. The experimental embryo has in addition to the normal cross-sectional complement of cells, extending to the arrowheads, an ectopic midventral strip of somite in continuity with its ventral mesoderm, as well as a neural formation, extending along much of the body. Proportion of host mesoderm devoted to somite is specifically reduced in these regions.

A type of operation, the heterochronic graft, where a dorso-lateral stripe of neurula mesodermal mantle together with overlying neuroectoderm and posterior zone of recruitment at its blasopore margin, is grafted to a narrow mid-ventral slot in a younger gastrula stage sibling. Zones of mesodermal recruitment and neurectodermal layer are matched up in graft and host and heal into continuity. The developing mesodermal cylinder now has, in addition to an approximately normal-sized host-derived cell population, a mid-ventral strip of older tissue which proves to have maintained its advanced schedule of development and to be in fact determined, under these conditions, to give rise to somite alone or to somite flanked on one side by pro-nephric differentiations. The notochord territory has been excluded from grafts and none of this structure in fact develops, but a small neural plate or rodlike neural formation usually overlies the ectopic somite, since neural fold was part of the graft.

Summarizes in histogram form the results of quantitative pattern analysis (see earlier section) in experimental *Xenopus embryos* of this type, in relation to their sham-operated control siblings. Total cross-sectional cell number in *host* mesoderm is not significantly different from normal, but the proportion devoted to somite is quite significantly reduced, the effect being very variable but present in most embryos. This somite deficit is largely concentrated in those parts of the axis lying opposite the relatively posterior position of the ventral ectopic somite in the mesoderm. Pronephros proportions are much more variable among experimentals than in controls and, on average, somewhat lower. But analysis of individuals reveals that, in those experimental embryos where pronephric proportions are diminished right below the normal range of variation in controls, a pronounced pro-nephric differentiation has occurred in graft-derived structures ventrally. Axolotl embryos behave similarly in response to these heterochronic grafts, though there the specific response of somite-proportion is distributed more widely along the host's axis.

These results make it quite improbable, in my view, that a positional information principle underlies control of the medio-lateral dimension of the body plan. The alternative theory proposed here would, however, accommodate and even predict

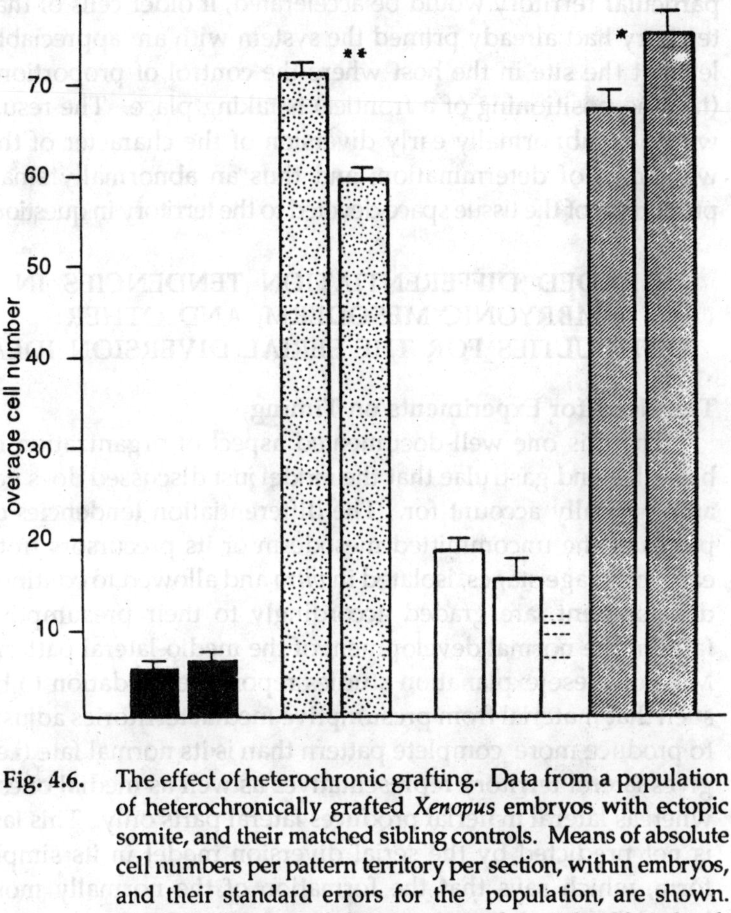

Fig. 4.6. The effect of heterochronic grafting. Data from a population of heterochronically grafted *Xenopus* embryos with ectopic somite, and their matched sibling controls. Means of absolute cell numbers per pattern territory per section, within embryos, and their standard errors for the population, are shown. Control columns on left, experimental on right. Notochord, solid black; Somite, stippled; Pronephros, clear; Pronephros in four individuals showing ectopic graft-derived pronephric differentiations, dashed index lines; Lateral plate, hatched. Key. ** = difference significant at p0.002, * = difference significant at p0.05.

them. They would represent feedback effects exerted by specific signals of the sort that normally regulate the pattern, but in this case emanating from advanced representatives of the territories concerned, that have been joined to the cell sheet at one extreme of the pattern. Buildup of signal specific to a particular territory would be accelerated, if older cells of that territory had already primed the system with an appreciable level at the site in the host where the control of proportions (i.e., the positioning of a frontier) is taking place. The result would be abnormally early diversion of the character of the wavefront of determination, and thus an abnormally small proportion of the tissue space devoted to the territory in question.

GRADED DIFFERENTIATION TENDENCIES IN EMBRYONIC MESODERM, AND OTHER DIFFICULTIES FOR THE SERIAL DIVERSION IDEA

The Need for Experiments on Timing

There is one well-documented aspect of organization in blastulae and gastrulae that the model just discussed does not automatically account for. The differentiation tendencies of pieces of the uncommitted mesoderm or its precursors from early cleavage stages, isolated in *vitro* and allowed to continue development, are graded accordingly to their presumptive fates in the normal development of the medio-lateral pattern. Most of these explanation studies report the gradation to be such that material from presumptive medial territories adjusts to produce more complete pattern than is its normal fate (i.e., gives lateral territory representatives as well as medial ones), whereas lateral material produces lateral parts only. This last is not predicted by the serial diversion model in its simple form, which says that the formation of the normally more lateral differentiations is consequent on the prior development by other cells in the system towards earlier, normally more medial ones. The expectation is thus that all isolates in which differentiations are achieved will preferentially form representatives of medial territories, and that more lateral ones will only be achieved insofar as each explant maintains or reachieves a spatial gradation in intrinsic rates of maturation within its material, such as is normally set up by communication with the organizer. But differentiation patterns in explants can be produced without notochord. Furthermore, under particular conditions in whole embryos, notochordless body patterns can be found that are otherwise normally organized in the medio-lateral dimension. There is no reason, in such cases, to postulate the death of cells that would otherwise have become notochord; such cells have presumably contributed to the somite that now marks the medial extreme of pattern.

From limited published work and from informal exchange of information, it appears that the capacity for differentiation as isolated explants develops earlier in the mesodermal precursors of embryos of the frog type than in those of salamander type, and that its dorso-ventrally graded properties when explanted are somewhat less pronounced in the former type. This is reminiscent of another quantitative difference in the progress of pattern control as between the two types. If a salamander-type embryo is divided into a smaller dorsal and a larger ventral fragment when gastrulation activity has already started in the dorsal midline (the organizer), the dorsal fragment can continue development and regulate its fate to form pattern to a considerable degree. The ventral fragment however, although surviving for a long time, usually differentiates little except epidermis, yolky endoderm, and, after prolonged delay, a primitive lateral platelike mesenchyme with blood islands. In frog embryos subjected to the same procedure, the ventral fragment has already acquired the capacity to gastrulate and to develop mesodermal pattern to a large degree, though the medial extreme of pattern is missing.

The general trend is for all these partial patterns deficient in medial territories, however produced, to result from abnormally slow schedules of development and (presumably) determination in the cells. These are associated either with the early treatment of the whole embryo, or with the ambient conditions following grafting operations or the culture of isolates. Proper quantitative study of the time schedules associated with the development of all examples of abnormal patterns will be of prime importance in assimilating all the data into a coherent theory for pattern control in this system. However, the following outline might accommodate both the concept of serial diversion arrived at in a previous section and the phenomenon of patterns without medial parts mentioned above. The effect of the organizer is assumed to be one of "spreading activation," propagating into surrounding tissues at sometime before its first expression as the spreading of gastrulation activity to each location. Such activation of progress towards gastrulation and the determined stage may start sooner and progress relatively more rapidly in frog-type than in salamander-type embryos, hence the systematically less graded character of development in frog isolates, and the less delayed and incomplete morphogenesis of the ventral fragments in these embryos. The degree of activation achieved locally may steeply affect the rate of cellular development, ensuring the gradient rate needed to effect a wavefront of determination susceptible to serial diversion. But such an organization would also mean that in cases of subnormal activation or attenuation of the apical (organizer) end of the activation gradient due to specific early treatments, or perhaps loss of activation through early isolation in culture or tissue disruption , development is slowed to varying extents. The more "medial" determined states may become sequentially unavailable with prolonged passage of physiological time in cells, even in the absence of diverting signals from "older" cells elsewhere. Total isolation from activation may lead always to blood island/lateral plate development after a prolonged period. In short, the normal pattern developed in *vivo* may be the combined outcome of a (nonregulatable) gradient system in degree of activation, itself biasing the "first choice" state of determination at each location, together with the series of specific diverting signals which ensure proportionality and which prohibit the formation of *laterally* incomplete patterns in small and in organizer-grafted embryos. The specific signals would also underlie the results of heterochronic grafting.

The gastrulation movements themselves occur according to a pronounced graded time schedule that appears quite well controlled as between individuals. Presumptive blood island cells begins their immigration a set time after the presumptive notochord (organizer) region. The precise relation between the endogenous controls of rate in these early activities and those controlling onset of determination for pattern is unknown, but it will be of great relevance for the theories so far discussed to time the visible dorso-ventral gastrulation sequence in small and in organizer grafted double dorsal embryos in relation to their normal siblings. embryos of both types are able to reduce the sizes of territories so as to complete pattern in the spaces available. Regulation of a true positional signal gradient, if successful, might be expressed in the gastrula stage as normal schedules of progression in absolute time, spanning the reduced dorso-ventral tissue spaces available. That is, little gastrulae and the two half fields of double gastrulae should each take a *normal* time for immigration activity to proceed across them from its onset mid-dorsally. By contrast, the primitive sort of activation system, necessary for pattern coherence on the serial diversion model, might be expected to behave like the profiles number 1 or 2. If timing reflected the gradient profile, this would lead to a contrasting phenomenon at gastrulation; the completion of the blastoporal ring by rapid spread of the immigration activity in less than normal time. These simple, relevant but exacting experimental observations remain to be made.

The units of pattern also called "territories" do not correspond simply to determinations for development into cell types as recognized by the cytologist or the cell physiologist. They are, rather, particular geographical units where cells' possible development becomes restricted to formation of particular components of the final body, each component then contributing a multiplicity of cell types and structures. Somite contributes skeletal and muscle tissue, while lateral plate and blood-forming islands contribute an even greater variety of histological compartments in the finally differentiated frog. There are thus limits to an analogy between the concept of pattern formation by serial diversion of determination due to specific signals, and that of control over the functional sizes of different tissue systems in the later body by systemic "chalones" (e.g., erythropoietin, liver-chalone). At the time of pattern control, the embryo has no vascular system, but the cells of the entire mesoderm are probably in communication *via* gap functions allowing passage of molecules up to around 1,000 molecular weight, and *via* the restricted extracellular diffusion space between the sealed faces of ectoderm and endoderm. The postulated signals are thus more allied to traditional morphogens than to hormones or chalones, and could conceivably be their common evolutionary precursors.

THE ANTERO-POSTERIOR DIMENSION

The Regulation of Longitudinal Pattern

Pattern in the antero-posterior or cephalo-caudal dimension of the body plan is also subject to a morphallactic regulatory process at early stages, at least as regards anterior regions down to the level of the hind limb rudiments. Experimental embryos of the type discussed in an early section with medio-lateral pattern down to below half-normal scale, also show normal numbers of abnormally short somites (mesodermal segments) between unique markers such as ear vesicle and root of the tailbud. There is reduced cell number in the original long dimension of each somite at its formation, so that a normal complement of these structures can be made with the tissue available. These embryos also show reduced cell number in the long dimensions of the pronephros, the heart and the brain parts, the last presumably a reflection of shorter zones of inductive activity in the mesoderm.

In order to normalize longitudinal pattern to small size in this way, embryos must heal after size reduction by middle blastula stages. Transection of later gastrulae and neurulae in the transverse plane, or transposition of somitogenic tissue along the axis at these times, results in highly mosaic behaviour. That is, pieces of embryos isolated or transposed at such stages differentiate, with respect to numbers of somites developed in each piece and the timing of their segmentation, like pieces of a photographic film exposed to a complex scene, then cut up, then developed. It therefore appears that although the visible processes demarcating successive segments from the somitogenic tissue occur in mesoderm which has been in its definitive position for sometime, the original factors which scale the processes and which set their local timing to give a "wavefront" progressing down the body plan, must be subject to a regulatory process acting before or at gastrulation. The result of this early process must be some graded cell state along the presumptive axis, which sets the smoothly graded time-sequence for local events as well as the different determinations attained, and which transits the same absolute range of values from anterior to posterior pattern limits in normal and in small embryos. A system which can do this is indeed reminiscent of true positional information, if in fact the cells that are to make the pattern are laid out in coherent sequence in a sheet at the time their *"values"* are given them. But our understanding of the fate map for this pattern dimension, whereby tissue is recruited into the mesoderm in its antero-posterior sequence during gastrulation, is not complete. Extensive cell mixing could conceivably occur with respect to position along this axis as between blastula and gastrula stages, though this seems unlikely.

Positional Signals, Nonequivalence and Timing

Most embryologists feel that we are presently much further from understanding the mechanism of antero-posterior pattern regulation than that of medio-lateral pattern regulation in the vertebrate embryo. The positional information concept still seems helpful for the longitudinal axis, especially if it is broadened to include the idea of positional value being given by a mechanism that measures time spent by cells in the marginal zone before their recruitment into the mesoderm, rather than their physical positions within a spatial gradient of a signal. It is along the antero-posterior axis that the phenomenon referred to by Lewis and Wolpert as non-equivalence is most evident in primary pattern formation. Thus, the medio-lateral dimension of pattern described earlier, with its zones, however, tissue undergoing similar initial behaviour and differentiations, e.g., as somite or lateral plate, can be shown to be encoded with unique properties at each level. The most striking work demonstrating this has been carried out on bird, rather than on amphibian embryos. The tissues' early non-equivalence is revealed by the autonomous nature of the overall developmental sequences carried out by regions when transplanted elsewhere along the axis, but also, most obviously, in the specific physiological "age" in development at which the morphogenesis takes place in tissue from each level. Thus pre-segmental material from the neck somite region, transplanted to the thoracic or lumbar region, will ultimately produce there vertebrate of "neck" character. Similarly, lateral plate mesenchyme of fore- and hind-limb regions acquires very early its specific character, determinative of the type of limb produced when ectodermal jackets and mesoderms are cross combined.

Time, or rate as a continuous variable, seems to be fundamental to the expression and perhaps the genesis of pattern in the longitudinal axis. If experimentally small amphibian embryos having undergone morphallactic regulation are synchronized precisely with normal siblings at the early neurula stage (i.e., at the anterior onset of somite segmentation), their subsequent progress is parallel in terms of *numbers* of somites segmented and thus proportion of the axial pattern developed, rather than in terms of absolute cell numbers incorporated into somites and other axial structures. The latter arrangement would complete the development, by using up tissue, earlier in the smaller embryos, whereas in reality the same absolute time span is linked to the development of the complete pattern in both cases. Taken overall, the evidence is for a great multiplicity of states coded into tissue to give rise to this dimension of pattern. Such a multiplicity could not readily be regulated by a series of unique size-assessment/switching events such as were postulated for the relatively simple mediolateral dimension of pattern. Instead, some property occupying a graded range between absolute boundary values is implied, as in positional information theory, although there are problems in considering how such information might be used in controlling numbers of similar structures.

A Note on Universality

There is no a priority reason why control of the body pattern in the mesodermal cell layer should not be achieved by different mechanisms in each of its two dimensions. Frankel has suggested that the regulation of cortical geometry in ciliated protozoans has such a dual character, as well as drawing out the parallel phenomena of organization as between this and pattern regulation in the early development of vertebrate embryos. A similar dualism may apply to pattern establishment in the insect blastoderm. In another early episode of vertebrate development, namely the establishment of antero-posterior limb pattern, the data are currently consistent with the idea of a true, interpreted positional gradient underlying the character of the skeletal elements. There is even evidence in the chick limb-bud to suggest that by contrast with the primary body pattern discussed in this chapter, growth control is correlated with pattern control so that the gradient mechanism for the latter may be simple and primitive, and yet avoid the difficulty of regulation for size because size is itself controlled accurately at an early stage. Should this prove to be the case, the way is open to reconsider yet other concepts for control of the relatively complex pattern that characterizes, say, the limb skeleton. These concepts come under the general heading of *"pre-patterns."* Unlike positional information, or the serial diversion model outlined here, pre-patterns involve the concept of a spatial distribution in the morphogen levels that is isomorphic with the future pattern, i.e., a set of peaks and troughs of concentration, rather than being a simple gradient that requires a complex genetic response from the cells.

A problem with the physico-chemical model systems whereby such pre-patterns might emerge, has always been that the pattern produced is crucially dependent upon the scale of the system, for reasons similar to those whereby reaction-diffusion-controlled gradient profiles are similarly dependent. If limb rudiments in fact contrive to keep spatial scale constant during the initial specification of pattern within them, then some (though not all) aspects of these somewhat complex patterns may depend upon pre-patterning processes.

The possibility is emerging that, rather than a single universal form of process, there may be a selection from a small repertoire of available processes according to functional considerations connected with particular dimensions of pattern in particular embryos at particular times. The various classes of process touched on in this chapter are all, nevertheless, to be thought of as morphallactic if indeed the term is to continue in modern use. In none of them is the development of a particular spatial array of cells with specific potentialities intimately tied, *via* neighbour interactions, to the process of tissue production by cell division itself.

5

Regeneration in Vertebrate Limbs

The vertebrate limb had long been the centre of the debate leading up to the publication of the *Origin of species* and there seemed to be a common plan to the structure of all vertebrate limbs. The term coined to describe such similarities was *'homology'*. Much was made of the homology of vertebrate limbs by Darwin and the supposed explanation of the phenomenon was that the limbs of each species had descended from a common ancestor whose archetypal limb pattern had subsequently under-gone modification. Despite the fact that these conclusions were derived from studies of the anatomy of adult limbs, as many others have been since then, the common ancestor theory had implications for embryology. First, it became accepted that all vertebrate limbs passed through this common ancestor stage in their development, without any embryological evidence to support it. Only recently have detailed studies been performed to discredit this view and it is now clear that the prechondrogenic condensation pattern of the limb bud mirrors, in miniature, the pattern of the adult form. The limb of each species therefore develops uniquely in terms of the prechondrogenic condensation pattern, a conclusion which is reinforced below.

Secondly, in evolutionary terms, development is viewed as a highly ordered sequence of hierarchical processes. This means that the earliest events in the development of the limb should be common to all vertebrates, with adaptation only working on the latest stages to provide the variety of adult forms we see. In this hierarchical scheme, one of the earliest processes to occur in limb development must surely be the establishment of the limb axes, without which organised growth could not take place. All vertebrate limbs have an antero-posterior, a dorso-ventral and a proximo-distal axis, and, according to the implications of the above evolutionary thinking, each axis should be organised by the same mechanisms throughout the vertebrates.

We do not know even today the precise details of how the axes of developing limbs are organised. The only techniques presently available to us to gain this knowledge are the observation of normal development and experienced interference in various ways. By performing these experiments on several species of embryo we can test the above hypothesis. Clearly, the implications are that, if all vertebrate limbs organise their axes by the same mechanisms, the effects of experimental interference should be the same. There have up to now been almost no studies of what might be called experimental homology, since most experimenters prefer to concentrate their efforts on the embryos of only one species. The beginnings of this study were reported recently and the work described here continues this effort. Four species of vertebrate limb are compared — the developing limb of *Ambystoma mexicanum* (urodele amphibian), *Rana temporaria*, *Xenopus laevis* (anuran amphibian) and *Gallus domesticus (Aves)* as well as data on the regenerating limbs of *Ambystoma mexicanum*. We can therefore look for similarities not only between the developing limbs of amphibians and birds, but also between developing and regenerating limbs.

Normal Development

To observe the pattern of emergence of prechondrogenic elements limb buds at various stages of development can be stained with Alcian green. We find that this pattern reflects faithfully the cartilages in the adult form. There are no similarities between the stages of developing limbs of axolotl, *Rana, Xenopus* or *chick*, indeed the elements do not even develop in the same antero-posterior sequence. For example, in axolotl limb buds the digits appear in approximately an anterior to posterior sequence — in the hindlimb the sequence is digits 2, 1, 3, 4, 5 but in *Rana* and *Xenopus* the sequence is from posterior to anterior — digits 4, 3, 5, 2, 1. In the chick, where the forelimb bud is routinely used, the sequence is 4, 3, 2. Furthermore, the chick limb bud condensation pattern does not pass through a primitive amphibian stage as the common ancestor theory would have us believe. A more detailed examination of the size of elements when they first appear in the developing limb has also revealed differences between species. In chicks, the size of the elements in each of the four segments (humerus, radius and ulna, wrist digits) when they appear in the proximo-distal sequence is approximately 300 µm. We might expect such a fundamental mechanism to be common to all vertebrates, but again this seems not to be so — each species has its own characteristics element size. In axolotls, the figure is approximately 250 µm, in *Xenopus* 150 µm and in *Rana 130 µm.*

Work on the Proximo-distal Axis

For the present study we can perform two types of operation. In the first, the tip of the limb bud or regeneration blastema is simply cut off, leaving about 50% of the original amount of tissue. Can the remaining tissue regulate to form a complete limb? The result of this operation on either developing or regenerating limbs of axolotls is the production of perfect limbs in 100% of the cases. In *Rana* limb buds, a few limbs with defects begin to appear -95% of limbs regulate. In *Xenopus*, only 38% of limbs are perfect and, in chicks, all limbs are truncated at the level of the cut. Thus a clear decrease in the regulative ability is apparent, as it is in the second type of experiment. Here, the distal tip of the limb bud or regeneration blastema is cut off and a second cut made at a more proximal level. This slice of tissue is removed and the distal tip replaced on the proximal stump. We can then ask whether the intercalary defect is regulated. Again, in axolotl limb buds and regeneration

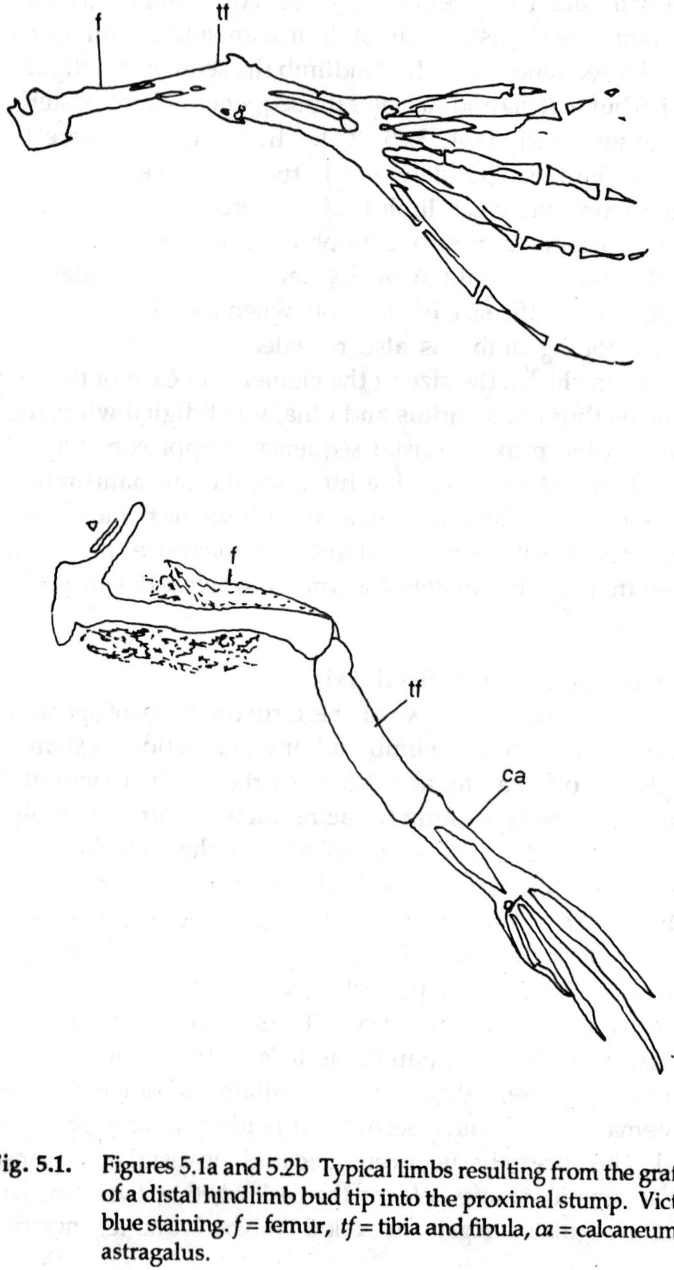

Fig. 5.1. Figures 5.1a and 5.2b Typical limbs resulting from the grafting of a distal hindlimb bud tip into the proximal stump. Victoria blue staining. *f* = femur, *tf* = tibia and fibula, *ca* = calcaneumand astragalus.

blastemas, perfect regulation takes place and 100% of the limbs are normal. In *Rana* limb buds, this operation only results in 42% of perfect limbs, and in *Xenopus* and chicks all limbs have deficiencies of one kind or another, although the results are highly stage dependent. Thus again a decreasing regulative ability is apparent throughout these vertebrate species.

TABLE 5.1

Frequency of Supernumerary Limb Induction after Three Types of Operation on the Developing Limb Buds of Rana, Xenopus and Chick and the Regenerating Limb Blastemas of Axolotl

Animal	*Operation*	*Number of cases*	*Number of limbs with supers (%)*	*Number of supernumeraries* Single (%)	Double (%)	Triple (%)
Rana	AP inversion	26	54	21	79	0
Xenopus	AP inversion	37	30	90	10	0
Chick	AP inversion	32	31	90	10	0
Axolotl	AP inversion	143	57	81	19	0
Rana	DV inversion	28	61	12	88	0
Xenopus	DV inversion	34	26	89	11	0
Chick	DV inversion	18	0	0	0	0
Axolotl	DV inversion	37	81	81	19	0
Rana	AP/DV inversion	58	67	46	41	13
Xenopus	AP/DV inversion	43	47	65	30	5
Chick	AP/DV inversion	31	68	76	24	0
Axolotl	AP/DV inversion	83	94	57	40	3

AP = antero-posterior; DV = dorso-ventral.

Work on the Transverse Axes

Investigations of the organisation of the transverse axes (antero-posterior and dorso-ventral axes) have involved the grafting of limb buds or regeneration blastemas to invert either one or both axes. By grafting a left limb bud to a right stump (or vice versa) each axis can be inverted independently, and by grafting a limb bud back onto its own stump after 180° rotation both transverse axes are inverted at once.

The Antero-posterior Axis

Inversion of the antero-posterior (AP) axis results in the production of *supernumerary limbs,* their frequency varying according to the species concerned. They arise at the points of

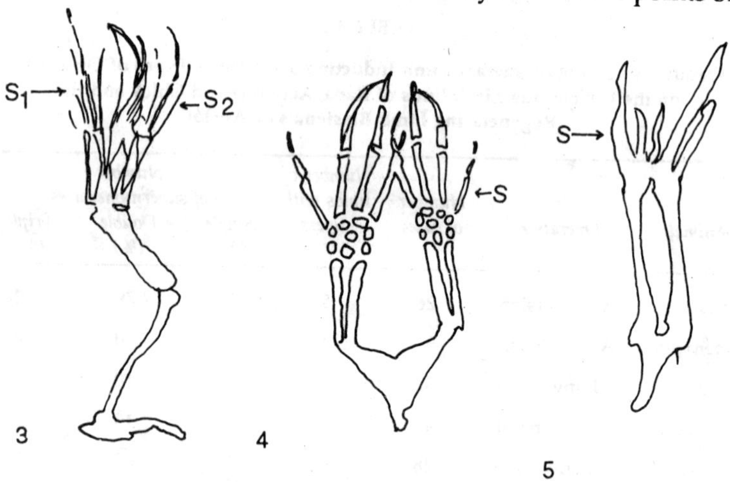

Fig. 5.2. Supernumerary limbs produced by contralateral grafting of limb buds of regeneration blastemas to invert the antero-posterior axis. Victoria blue or Alcian green staining. Rana hindlimb with two supernumeraries S_1 and S_2 at the anterior and posterior poles of the limb. In between is the grafted contralateral limb. The digital sequence from left to right is 1 2 3 4 5 (S_1), 5 4 3 2 1 (Graft) 1 2 3 4 5 (S_2). Note the supernumeraries are in mirror-image relation to the grafted limb. A single supernumerary from an axolotl regeneration blastema graft. On the left is the grafted limb with a digital sequence of 4 3 2 1 and on the right is the supernumerary (S) with a digital sequence of 1 2 3 4 in mirror-image relationship. A single supernumerary (S) produced by grafting a chick limb bud. The similar mirror-image digital sequence of 4 3 2 2 3 4 apparent here too.

maximum incongruity, i.e., at the anterior and/or posterior poles, and can be either single or double. The relatively low incidence of supernumerary limbs produced by AP inversion in chick limbs should be contrasted with the effect of grafting the organising region from the posterior margin of the limb bud to the anterior margin. Using equivalent stage embryos, supernumeraries can be typically produced in more than 90% of the cases.

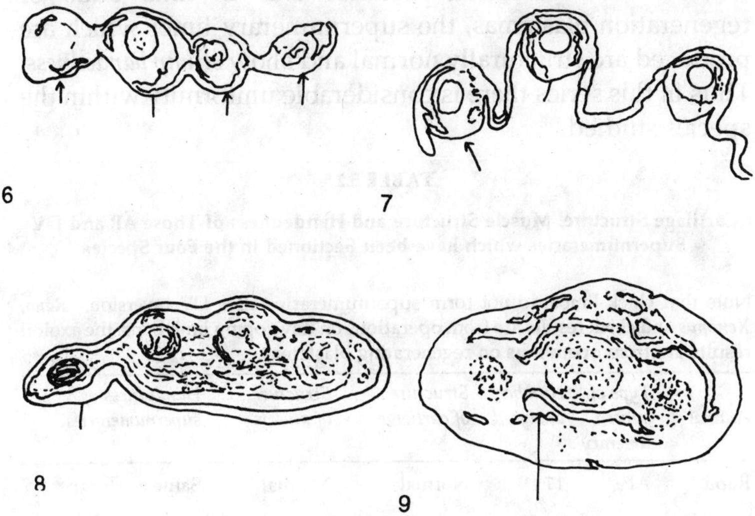

Fig. 5.3. Sections through the digits of normal limbs showing the dorso-ventral asymmetry of muscle patterns. Haematoxylin and eosin staining.

Analysis of cleared whole mounts reveals entirely normal sequences of digits, carpals etc. This is, the AP axes of the supernumeraries follow a well-defined pattern, even if a full complement of digits is not present. In Figure for instance, the sequence of digits is 1, 2, 3, 4, 5 (supernumerary) 5, 4, 3, 2, 1 (graft) 1, 2, 3, 4, 5 (supernumerary), and the sequence is 4, 3, 2 (supernumerary) 2, 3, 4 (graft). To determine the organisation of the dorso-ventral (DV) axis, AP supernumeraries have been serially sectioned to study their muscle patterns and together these two methods of analysis allow one to determine their handedness. The normal muscle structure for each species is depicted. It is characterised in the amphibians by the early disappearance of dorsal muscles and the persistence of ventral muscles to the tips of the digits. The muscles of the chick, too, have a characteristic dorsoventrality. Analysis of total of 38 supernumeraries revealed that in every case their muscle patterns were entirely normal. From this we can conclude that, after inversion of the antero-posterior axis in limb buds or regeneration blastemas, the supernumerary limbs which are produced are structurally normal and show *stump handedness.* Thus in this series there is considerable uniformity within the species studied.

TABLE 5.2

Cartilage Structure, Muscle Structure and Handedness of Those AP and DV Supernumeraries which have been Sectioned in the Four Species

Note that chick limbs do not form supernumeraries after DV inversion. *Rana, Xenopus* and chick results are from operations on developing limb buds, the axolotl results are from operations on regenerating blastemas.

Animal	*Type of supern-umerary*	*Number analysed*	*Structure of cartilage*	*Structure of muscle*	*Handedness of supernumeraries*
Rana	AP	11	Normal	Normal	Same as stump
Xenopus	AP	6	Normal	Normal	Same as stump
Chick	AP	10	Normal	Normal	Same as stump
Axolotl	AP	11	Normal	Normal	Same as stump
Rana	DV	11	Normal	Normal	Same as stump
Xenopus	DV	6	Normal	Normal	Same as stump
Axolotl	DV	10	Normal	Normal	Same as stump

AP = antero-posterior ; DV = dorso-ventral.

The Dorso-ventral Axis

Inversion of the dorso-ventral axis results in the production of supernumeraries in the amphibian species, but not in the chick. In the former, one to two supernumerary limbs are produced at varying frequencies, usually at the points of maximum incongruity which, this time, are the dorsal and ventral poles of the host limb. Analysis of the muscle patterns in *Rana, Xenopus* and axolotl regenerating limbs revealed that, as in the previous case, all supernumeraries were normal in their dorso-ventral organisation as well as in their AP sequence of digits and were of *stump handedness.*

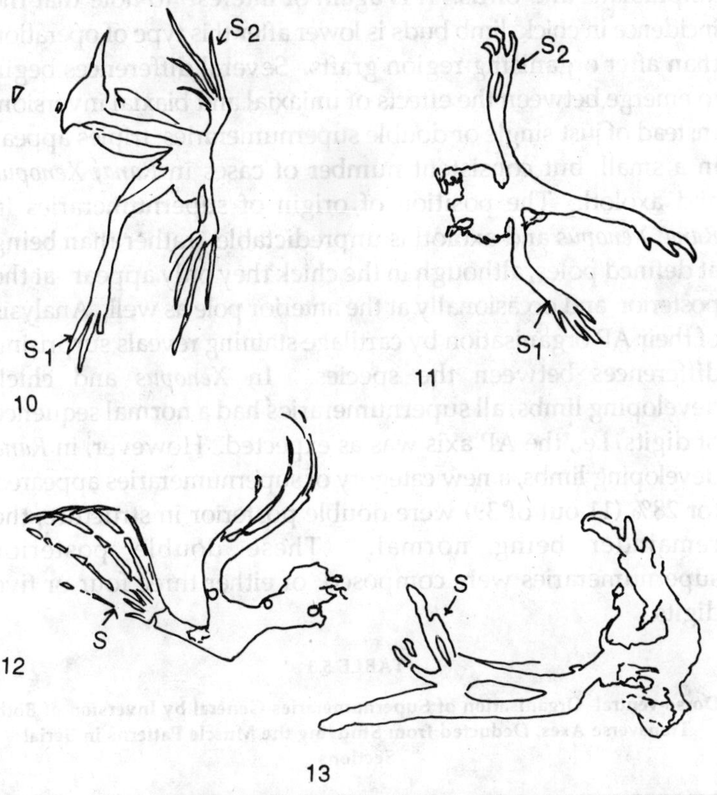

Fig. 5.4. Supernumerary limbs (S_1 and S_2) produced in a *Rana* limb after DV inversion of the grafted (middle) limb bud. The supernumeraries arose at the dorsal and ventral poles of the limb. Victoria blue staining. DV inversion of an axolotl regeneration blastema showing the same type of supernumeraries (S_1 and S_2) as the previous figure. Victoria blue staining. 180° rotation (AP/DV inversion) of a *Xenopus* limb bud which resulted in a single five-digit supernumerary (S). Victoria blue staining. AP/DV inversion of a chick limb bud which produced a single supernumerary. Alcian green staining.

Inversion of Both Axes

After 180° rotation of limb buds or regeneration blastemas, a high frequency of supernumerary limbs is produced in both amphibians and birds. It is again of interest to note that the incidence in chick limb buds is lower after this type of operation than after organising region grafts. Several differences begin to emerge between the effects of uniaxial and biaxial inversion. Instead of just single or double supernumeraries, triples appear in a small, but consistent number of cases in *Rana, Xenopus* and axolotl. The position of origin of supernumeraries in *Rana, Xenopus* and axolotl is unpredictable rather than being at defined poles, although in the chick they only appear at the posterior and occasionally at the anterior pole as well. Analysis of their AP organisation by cartilage staining reveals surprising differences between the species. In *Xenopus* and chick developing limbs, all supernumeraries had a normal sequence of digits, i.e., the AP axis was as expected. However, in *Rana* developing limbs, a new category of supernumeraries appeared for 28% (11 out of 39) were double posterior in structure, the remainder being normal. These double posterior supernumeraries were composed of either three four or five digits.

TABLE 5.3

Dorso-ventral Organisation of Supernumeraries General by Inversion of Both Transverse Axes, Deducted from Studying the Muscle Patterns in Serial Sections

Animal	*Numbertype of super-numeraries analysed*	*Numbers of each type of muscle pattern in AP/DV supernumeraries*			
		Normal	*2D or 2V*	*Part normal/ Part mirror-imaged*	*Part normal/ part inverted*
Rana	18	5	3	4	6
Xenopus	12	9	1	1	1
Chick	17	17	0	0	0
Axolotl	100	22	22	32	24

Examples of each type are — normal; double ventral (2V); double dorsal (2D); part normal/part mirror-imaged; part normal/part inverted. *Rana, Xenopus* and chick results are from operations of developing limb buds, the axolotl results are from operations on regenerating blastemas.

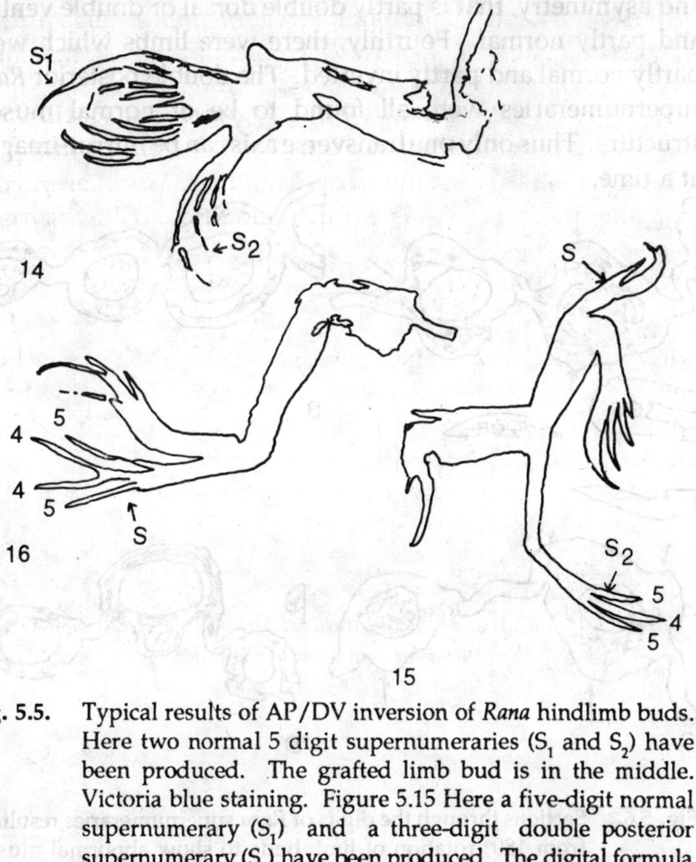

Fig. 5.5. Typical results of AP/DV inversion of *Rana* hindlimb buds. Here two normal 5 digit supernumeraries (S_1 and S_2) have been produced. The grafted limb bud is in the middle. Victoria blue staining. Figure 5.15 Here a five-digit normal supernumerary (S_1) and a three-digit double posterior supernumerary (S_2) have been produced. The digital formula of the double posterior limb is 5 4 5. Here a single four-digit double posterior supernumerary (S) with a digital formula of 5 4 4 5 has been produced. Above the supernumerary is the grafted limb.

Analysis of their dorso-ventral organisation by muscle patterns revealed further suprises. In axolotl regenerates and

Rana limb buds, there were approximately equal numbers of four types of muscle patterns. First, there were normal supernumeraries as had been found before. Secondly, there were perfectly mirror-imaged limbs, either double ventral or double dorsal. Thirdly, there were limbs of mixed symmetry and asymmetry, that is partly double dorsal or double ventral and partly normal. Fourthly, there were limbs which were partly normal and partly inverted. The double posterior *Rana* supernumeraries were all found to be of normal muscle structure. Thus only one transverse axis can be mirror-imaged at a time.

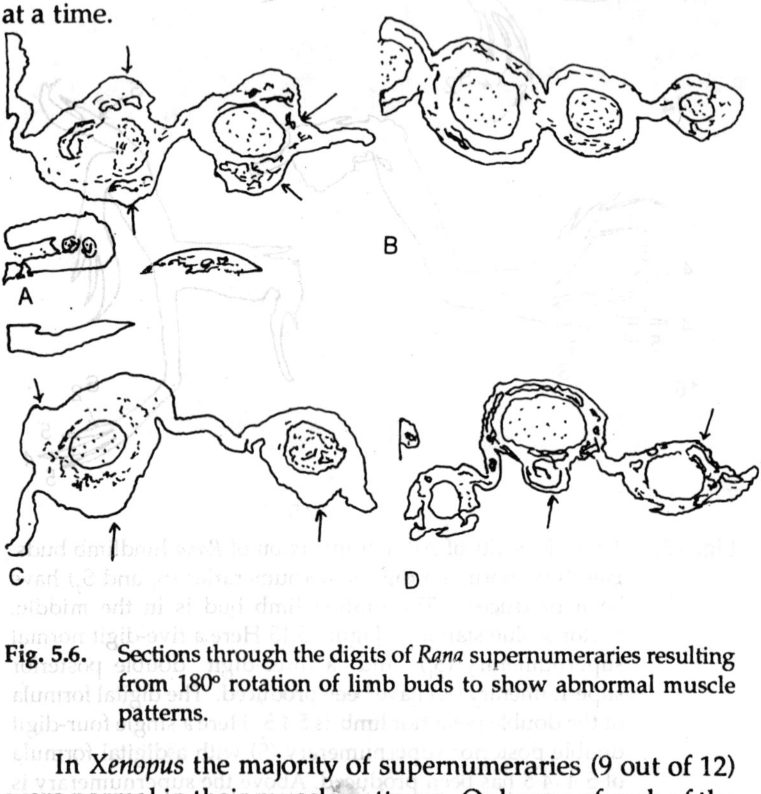

Fig. 5.6. Sections through the digits of *Rana* supernumeraries resulting from 180° rotation of limb buds to show abnormal muscle patterns.

In *Xenopus* the majority of supernumeraries (9 out of 12) were normal in their muscle patterns. Only one of each of the other three classes of structure was found. Supernumeraries produced in chick limbs often had a reduced complement of muscles, although those that could be identified were always arranged normally. Some effect on the dorso-ventral axis may be involved here, since AP supernumeraries had well-formed muscles.

Thus, after anteropostena/dorso-ventral inversion, there seems to be a gradation in the structure of supernumerary limbs generated from 100% normal (chicks) to various frequencies of four (*Xenopus* and *axolotl)* and ultimately five different classes of structure in *Rana,* including double posteriors.

The Effect of Vitamin A and Retinoids

One further method of analysing axial organisation which has recently come to light is the effect of vitamin A and its derivatives, the retinoids, on developing and regenerating limbs.

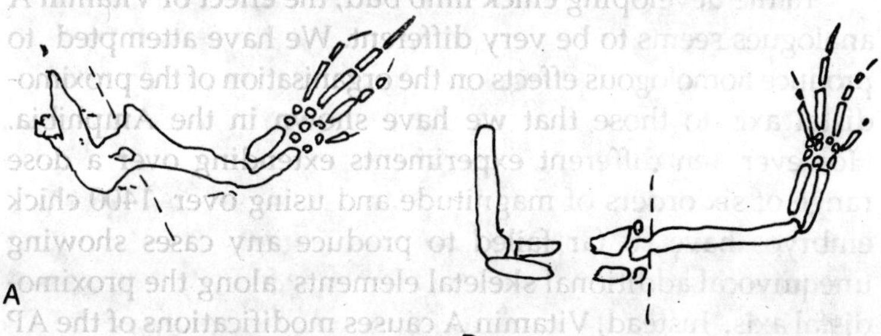

Fig. 5.7. Abnormal limbs resulting from Vitamin A treatment of regenerating axolotl limbs. Victoria blue staining.

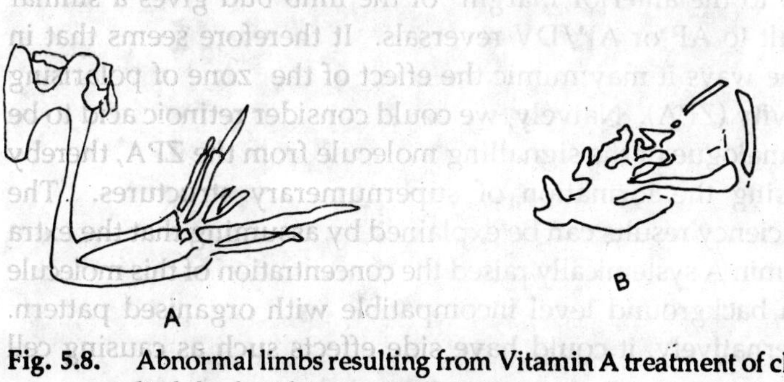

Fig. 5.8. Abnormal limbs resulting from Vitamin A treatment of chick limb buds. Alcian green staining.

If the amputated limbs of axolotls are immersed in a solution of retinol palmitate, then, instead of replacing those elements removed by amputation, extra tissue in the proximo-distal sequence is regenerated. For example, after amputating through the mid-radius and ulna, an extra radius and ulna can be produced in tandem as well as the hand. If the concentration of vitamin A is increased, a whole limb can be regenerated from the cut stump. The same phenomenon occurs more dramatically after amputation through the hand. It is important to note that, in both these figures, the antero-postena sequence of digits is normal, as is the dorso-ventral organisation of muscles when such limbs are sectioned. The effects of various retinoids have been rested and retinoic acid was found to be the most potent.

In the developing chick limb bud, the effect of Vitamin A analogues seems to be very different. We have attempted to produce homologous effects on the organisation of the proximo-distal axis to those that we have shown in the Amphibia. However, ten different experiments extending over a dose range of six orders of magnitude and using over 1400 chick embryos have so far failed to produce any cases showing unequivocal additional skeletal elements along the proximo-distal axis. Instead, Vitamin A causes modifications of the AP axis including the formation of mirror-image reduplicate supernumerary limbs or deficiencies in which parts of the limb are deleted. In particular, local application of retinoic acid to the anterior margin of the limb bud gives a similar result to AP or AP/DV reversals. It therefore seems that in some ways it may mimic the effect of the zone of polarising activity (ZPA). Naively, we could consider retinoic acid to be an analogue of the signalling molecule from the ZPA, thereby causing the formation of supernumerary structures. The deficiency results can be explained by assuming that the extra vitamin A systemically raised the concentration of this molecule to a background level incompatible with organised pattern. Alternatively, it could have side effects such as causing cell death, the inhibition of proliferation or the inhibition of cell differentiation.

Discussion

It is clear from the work described above that the four species of vertebrate limbs used here (three amphibians and one bird) behave differently in some aspects of development and similarly in others. The similarities, far fewer in number, seem to be only two. First, all limbs develop in a proximal to distal sequence of elements (stylopodium, zeugopodium, autopodium) rather than, for example, distal to proximal or various combinations of the two. Secondly, after inversion of the AP axis limbs all seem to develop supernumeraries of normal cartilage and muscle patterns with the same handedness as the stump.

On the other hand, the differences are many. During normal development the digits appear in an anterior to posterior sequence in the axolotl limb bud and in a posterior to anterior sequence in anurans and birds. The size of the elements when they first appear is characteristic of the species, as is the number and disposition of the elements. Experiments revealed that there is an increasing mosaicism in the proximo-distal axis from the axolotl to the chick which results in the loss of the ability to replace missing tissue. After inversion of the DV axis, amphibians produce supernumerary limbs of stump handedness, but chicks do not. After inversion of both transverse axes there are five types of supernumerary limb structure that appear in *Rana* limb buds, four in *Xenopus* and axolotl, and only one (normal) in chicks. Thus, the normal pattern seems to be most stable in the higher vertebrate class. Other differences between these limbs have also been observed such as the presence of an apical ectodermal ridge (AER), which plays an essential role in distal outgrowth, and its lack of regeneration in chicks. In anurans an AER is present and it is presumably capable of regeneration, but in urodeles there is *no* AER at all.

From this array of developmental behaviour in vertebrate limbs, we can draw one important conclusion. There seems to be more than one way of developing limb — it can involve highly regulative mechanisms such as those utilised by axolotl limb buds and regeneration blastemas, or it can involve mosaic mechanisms such as those in chick limb buds. The degree of mix of regulative and mosaic mechanisms seems to determine the developmental behaviour of the limbs of any one species. The conclusion is clearly contrary to the evolutionary view of the development of limbs, where such fundamental mechanisms as the organisation of the axes should be common to all vertebrates. Neither does there seems to be a common ancestral stage in development through which all limbs pass. Therefore not only have evolutionary concepts been of no value in advancing our understanding of developmental mechanisms, they have been positively misleading. There is surely no reason *a priori* why developmental mechanisms should not evolve and we concur with Sander (this volume) that homology does not imply conservation of mechanisms.

Finally, it is interesting to consider the relation between development and regeneration. Only urodeles are capable of limb regeneration in adults, and it is surely no coincidence that urodeles have the most highly regulative form of limb development. *Rana* can regenerate for a short period after limb development and have a less regulative form of limb development. *Xenopus* loses regenerative ability before limb development has terminated and has a relatively mosaic type of development. Chicks, which can never regenerate, are the most mosaic of all. It is possible therefore, that the reason why the limbs of higher vertebrates have not evolved regenerative ability is because their mosaic type of development precludes it.

6

The Evolutionary Development of Vertebrate Limbs

The phylogeny and the ontogeny of vertebrate limbs as peritadactyl limbs have been matters of considerable interest and dispute. Drawinians, reifying the abstract concept of a typical form, sought a common ancestral tetrapod limb structure in fossil amphibians such as *Eryops*, and antecedents in crossopterygian relatives such as *Sauripterus*, the limbs of later vertebrates being regarded as adaptive modifications of the ancestra form. Embryologists have looked for evidence of a basic pattern of cartilage condensation in the limb buds of developing amphibians, birds, and mammals prior to secondary modifications such as loss or fusion of carpal and tarsal elements. They were following the expectation, derived from the theory of recapitulation, that the ancestral form may appear transiently in the early ontogenesis of the limb. However, no consensus on an archetypal limb form has emerged.

This should not surprise us: the quest was initiated by an error of conceptualisation, the generalised form of which unfortunately runs through most of contemporary biology. This was Darwin's suggestion that the concept of the pentadactyl limb should not be treated as an abstract generalised form, as it was defined by Owen, but as an actual, materialised form which had a historical existence in an ancestor common to the line of terrestrial vertebrate descent. This is explanation of form by historical contingency, not by morphogenetic law, the later being the goal of the rational morphologists, such as

Geoffroy St. Hilaire, Cuvier, and Owen. The contemporary descendant of explanation by historical continuity and contigency is to be found in the concept that embryogenesis is a result of the operation of a genetic programme which is a historical record of adaptive success in a particular phylogenetic sequence. In this paper, we will not pursue this line of reasoning, but return instead to the concept that organismic morphology, and in particular vertebrate limbs, are to be understood in terms of general principles of spatial organisation, together with particular constraints which we develop in terms of the classical concept of the morphogenetic field.

Our objective is to demonstrate the simple proposition that tetrapod limb morphogenesis may be understood in terms of some basic generative principles capable of producing a great variety of limb forms which are all transformable one into the other under modifications of the limb-generating process. None of these forms is special in any way, none intrinsically more primitive' than any other, so that none can be selected as an ancestral form. Indeed, one of our purposes is to show that the study of principles or laws of biological form must be clearly separated from historical considerations, since the search for the invariants which define homology (equivalence of structure under a transformation which leaves certain spatial relations unchanged or invariant) depends solely on form and not on ancestry. Historical conservation of forms in particular lineages may also occur, but this observation does not help one to understand problems of biological organisation, such as the nature of the generative principles underlying the transformational equivalence of tetrapod limbs to which we now turn.

The Generalised Tetrapod Limb Form

The generalised tetrapod limb form is a primitive ancestral limb form, the so-called *'cheiropterygium'* illustrates the general structures and the relationships of their components to the different parts of the tetrapod limb. The Figure is from a contour plot produced by a computer programme based upon

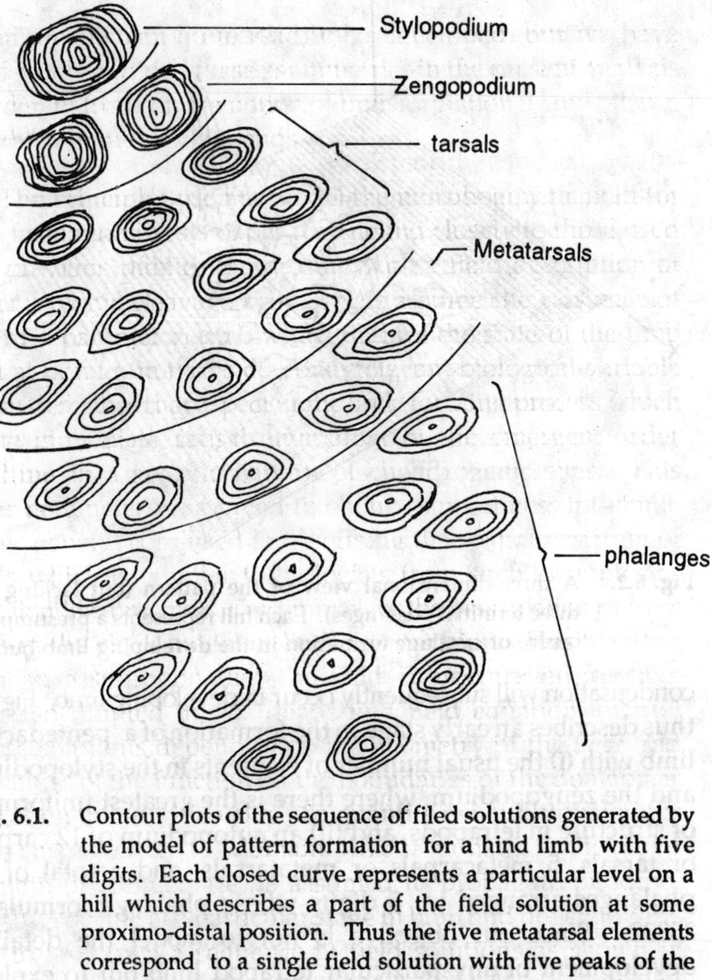

Fig. 6.1. Contour plots of the sequence of filed solutions generated by the model of pattern formation for a hind limb with five digits. Each closed curve represents a particular level on a hill which describes a part of the field solution at some proximo-distal position. Thus the five metatarsal elements correspond to a single field solution with five peaks of the function which is selected by a particular 'energy' level at this position in the proximo-distal axis.

a model of limb morphogenesis which is discussed briefly below and in detail in the Appendix. Each of the enclosed elements is a contour of a hill with a central maximum of a field function u(*x*), as shown in a three-dimensional perspective in Figures, and is taken to represent a domain where cartilage

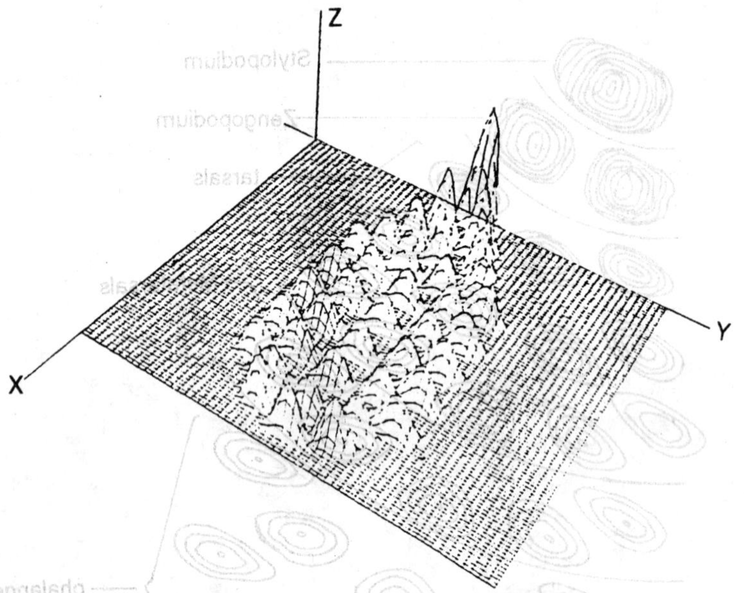

Fig. 6.2. A three-dimensional view of the pattern (but lacking the three terminal phalanges). Each hill represents a presumptive centre of cartilage formation in the developing limb bud.

condensation will subsequently occur in developing limb. Figure thus describes an early stage in the formation of a 'pentadactyl' limb with (i) the usual number of elements in the stylopodium and the zeugopodium, where there is the greatest uniformity of structure in tetrapods, and (ii) an autopodium of 12 carpals or tarsals, 5 metacarpals or metatarsals, and a total of 13 phalanges arranged in 5 digits whose phalange formula is 23332. Since our objective is not to model the detailed development of any particular tetrapod limb but to explore the generative principles common to this type of limb and thus to gain some insight into the concept of homology from a developmental perspective, we have concentrated on the proximo-distal sequence of limb elements and ignored the secondary modifications which occur in most limbs with respect to the antero-posterior axis. Thus all the limb patterns shown have antero-posterior symmetry. Models incorporating these spatial details can quite readily be developed, but we have deliberately avoided these asymmetries in the present analysis in order to focus on the notion of transformational equivalence on which homology depends.

The principles used to model the morphogenetic field for limb formation in this paper correspond closely to those used in a previous publication dealing with a field description of the embryonic cleavage process. We assume the existence of an order parameter, u(*x*), which defines the state of the limb field at point *x* in terms of some relevant biological variable which describes that aspect of the limb-forming process which is the immediate causal antecedent of the emergent order resulting in a specific pattern of chondrogenic sites. This order parameter is assumed to obey a 'smoothness' principle of the general type used in describing the spatial variation of fields which is familiar to biologists from such examples as diffusion-like processes 'sand-model' or the intercalation process used in the clock model of limb regeneration. This together with an assumption that the field values are everywhere bounded (limited in size) leads to a field equation for u(x) whose solutions depend upon the geometry of the field and the values of the function on the boundaries of the domain, as well as on the specific value of a quantity generally analogous to an energy within the domain. In our model, a semicircular 'domain of competence' is assumed to propagate proximo-distally, starting at a particular stage in limb bud development, and to lay down a sequence of field solutions as shown in Figure each antero-posterior array of elements being generated by a particular field solution within the domain. This assumption is similar to that used by Ede (1976) in his model of limb morphogenesis. Thus, the presumptive centre for the humerus or femur is laid down first by a solution that produces a single maximum of u(*x*) within the first hemispherical domain of competence. Then a solution with two maxima is established in a more distal domain generated by the advancing wave of competence; then a solution with three maxima, then four, then a series of four solutions with three maxima, then four, then a series of four solutions with five maxima, and finally a solution with three maxima are laid down in proximo-distal order. The particular sequence thus obtained is, in this model, dependent upon a proximo-distal gradient in the limb-bud which may be compared to 'energy' levels stabilising or selecting particular solutions of the field equation. Thus to get the solutions generated in Figure the gradient is assumed to decrease progressively form a value which selects the solution with one period, giving stylopodium, through values selecting solutions of periods 2, 3, 4 and 5, and finally rising to a value selecting the solution with period 3 again. The gradient is assumed to be continuous, since solutions are stable only at particular 'energy' levels in this type of model; i.e., we are dealing with an eigenvalue problem. Repetition of the same solution, such such as 5, occurs if the energy level is relatively constant so that solutions are set up as soon as a large enough domain is established by the advancing wave. To get the absence of a solution such as 4 between the distal 5 and 3 solutions, we assume a rather rapid change in the gradient so that a sólution of period 4 has insufficient space or time to get established as the wave of competence propagates distally.

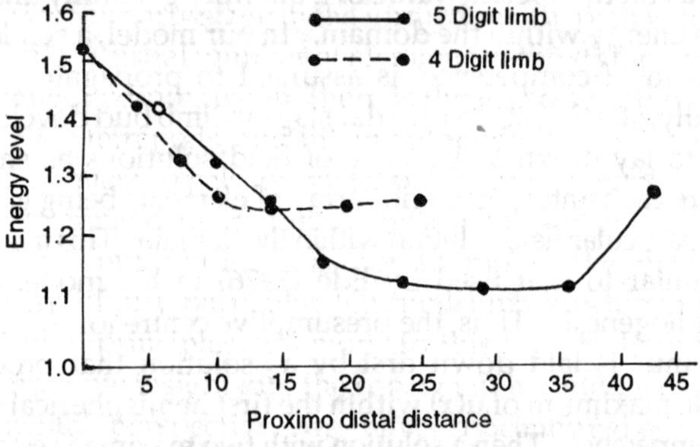

Fig. 6.3. The proximo-distal 'energy' profile which selects or stabilises a particular series of solutions of the field equation in successively more distal domains of competence in the limb bud.

The sequence in Figure corresponds to the 'energy' profile, which falls through a series of values generating the increasing series of elements, then levels off to give a sequence of four 5-unit solutions, and finally rises to give a terminal 3 unit solution. For comparison, a profile of a four-digit limb is included, which will be considered later.

Evidently it is the shape and the height of the proximo-distal 'energy' gradient which determine the sequence of elements which is laid down in a limb in this particular model, so that changes in the details of the gradient determine changes in pattern. This is only one of the possible ways of generating patterns by field solutions. A somewhat different mechanism is proposed by several workers, who whose diffusion-reactions equations for their fields and selected different solutions by means of changes in an effective diffusion constant. The similarity of the models is that both use solutions of the same field operator to give a prepattern defining where the skeletal elements will later appear in the limb. We could say of the present model that the proximo-distal wave of competence, together with the treatment of the process as an eigenvalue problems, thus giving quantised solutions across the antero-posterior axis, is what transforms a one-dimensional energy profile into a pattern which is periodic in two dimensions. A final point about the model is that no attempt is made to describe the modifications following initial chondrogenesis, such as the differential growth of elements in different parts of the limb or fusions of basipodial elements. Thus our concern is solely with the variety of forms which is possible in the primary condensation pattern, and how this may be generated.

Invariance and Variation

A simple modification involves a reduction in the total number of elements as a result of both a shorter proximo-distal series and a reduced periodicity across the antero-posterior axis. This pattern correlates with an altered 'energy' profile, (dashed curve). The level generating field solutions of period 5 is not reached, so the limb is one with only four

Y:	1.500							
R:	3.83	3.83	5.14	6.38	7.59	7.59	7.59	7.59
F:	0.70	0.71	0.74	0.78	0.81	0.82	0.82	0.82
A:	0.00	1.00	4.00	7.00	10.00	15.00	20.00	25.00
N:	1	1	2	3	4	4	4	4

Fig. 6.4. The limb pattern generated by the 'energy' profile of the dotted curve. *Y* is an arbitrary scale factor, *R*, *F*, and *A* are as described in the Appendix, while *N* describes the Bessel function Fm which has been selected by the energy value in accordance with equation (A_9). The first two Bessel functions corresponding to N = 1 at positions A = 0 and 1.0 no fuse to give the first chondrogenic site, after which the functions are separated.

digits. This is the type of limb found in urodeles such as the axolotl (*Ambystoma;* forelimb) or the salamander, *Necturus.* This drawing shows alternative ways in which a five-digit limb could undergo conversion to one with digits, by 'loss' of a digit either preaxially or postaxially. However, our generative model suggests a much simpler and more plausible solution: in a transformation from one limb form to another, there is no conservation of elements so that there can be no question of one digit being lost and others conserved. Biologists have been misled by this type of question ever since Darwin adopted a view of evolution which stressed the adaptive detail of organismic morphology (i.e., the particular ' content' of organisms) rather than principles of form, and problems of

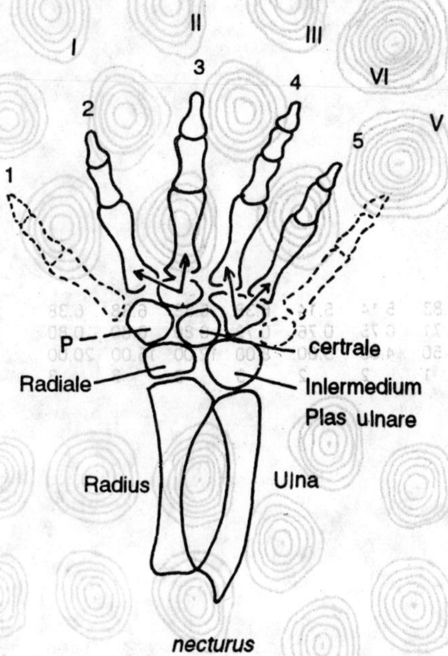

Fig. 6.5. Hypothetical alternatives in the evolution of a four-digit limb from a five-digit 'common ancestor', according to the Darwinian concepts of homology and inheritance. The original pentadactyl limb has digits 1-5, or I-V, and either digit I or digit V is then selected against to give a four-digit limb, 2-5 or I-IV, P = pisiform.

inheritance rather than those of organisation. Great attention has been paid to the exact number of phalanges per digit, their relative length, and other features which could identify their individuality and so allow them to be recognised as units of inheritance when they turn up in *Eryops*, in the salamander, in the wing of the domestic hen, or in some other organism with an even greater 'loss' of digits. Thus the three digits of the hen's wing are identified as digits 2, 3 and 4 of the ancestral five-digit limb, the two of the antelope are labelled 3 and 4, and the single toe remaining in the horse is identified as digit 3 Just how is such identification possible, especially when so much variation can occur in the details of the bone pattern

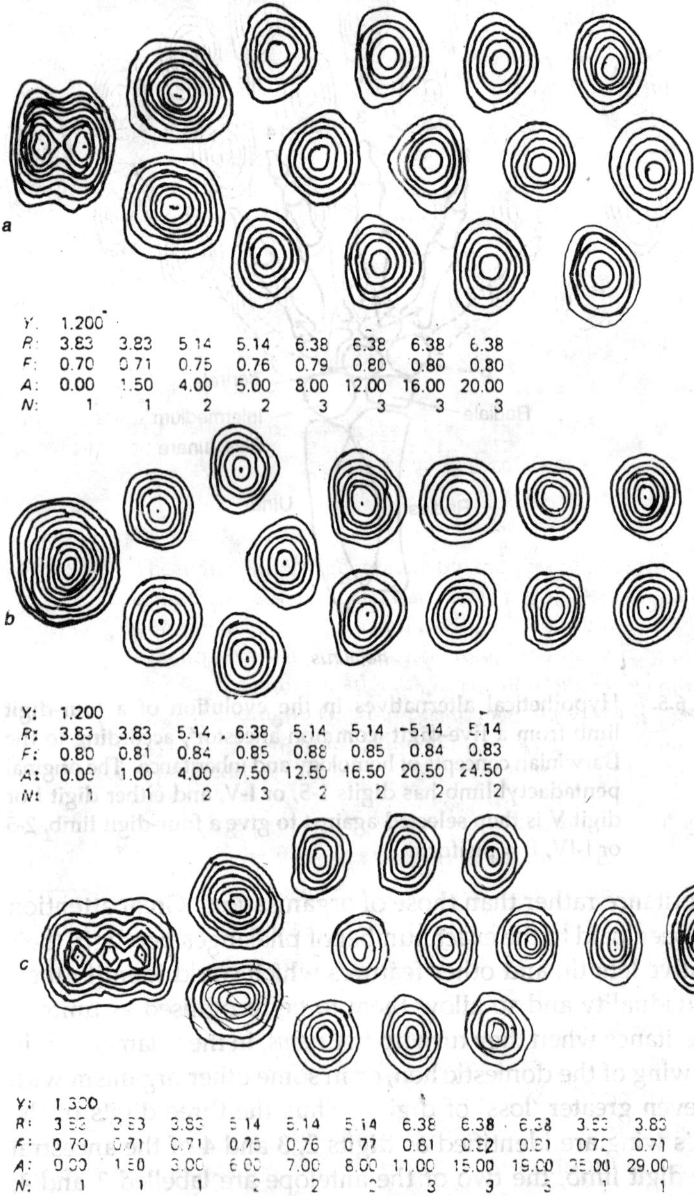

a

Y:	1.200							
R:	3.83	3.83	5.14	5.14	6.38	6.38	6.38	6.38
F:	0.70	0.71	0.75	0.76	0.79	0.80	0.80	0.80
A:	0.00	1.50	4.00	5.00	8.00	12.00	16.00	20.00
N:	1	1	2	2	3	3	3	3

b

Y:	1.200							
R:	3.83	3.83	5.14	6.38	5.14	5.14	5.14	5.14
F:	0.80	0.81	0.84	0.85	0.88	0.85	0.84	0.83
A:	0.00	1.00	4.00	7.50	12.50	16.50	20.50	24.50
N:	1	1	2	3	2	2	2	2

c

Y:	1.300										
R:	3.83	3.83	3.83	5.14	5.14	5.14	6.38	6.38	6.38	3.83	3.83
F:	0.70	0.71	0.71	0.75	0.76	0.77	0.81	0.82	0.81	0.72	0.71
A:	0.00	1.50	3.00	6.00	7.00	8.00	11.00	15.00	19.00	25.00	29.00
N:	1	1	1	2	2	2	3	3	3	1	1

Fig. 6.6. Limb patterns generated by modifications in the proximo-distal 'energy' profile of the limb bud, corresponding to limbs with (a) three, (b) two, and (c) one digit.

within a single species, or even in an individual with multiple regenerates a question to which we shall turn in a moment. Examples of the general categories of limb reduction to 3, 2 and terminal digit, without any effort to reproduce the detailed pattern of any particular species. What emerges very clearly at this level of abstraction is just the point made by the pre-Darwinian rational morphologists: biological patterns must be seen and understood as wholes which undergo transformation, so that there is no identity of elements in different patterns. It is wrong even to suggest that the most proximal element of the limb is the 'same' in all tetrapods, even though the element in this position is nearly always single. What is the same is the generating principle of the total pattern, so that there are invariants of universals which are recognised as common to the whole set of forms. These forms are thus equivalent under transformation. Each element is a component of a total pattern, but has no inherited individuality in a genealogical sequence.

Embryological Transformation

The conclusion reached above suggests that it may be possible to perturb limb morphogenesis in a single vertebrate species in such a way that the basic skeletal patterns of other vertebrates are produced. This is because, if the only universal constraint is the generative mechanism of the presumptive chondrogenic pattern, then perturbations to the initial and the boundary conditions of the limb field should result in perturbations in the pattern even against a fixed genetic background, since in a field description genes cease to play the role of specific causes of form. They contribute to the particular conditions which stabilise or select a specific field solution from the possible set, or a specific sequence of solutions which generate the field morphology, but their influence can be over-ridden by other particular conditions. Because of the absence of inherited individuality in the units of the limb, we should not expect to be able to identify elements in the perturbed limbs, but only general categories of tetrapod pattern. It is

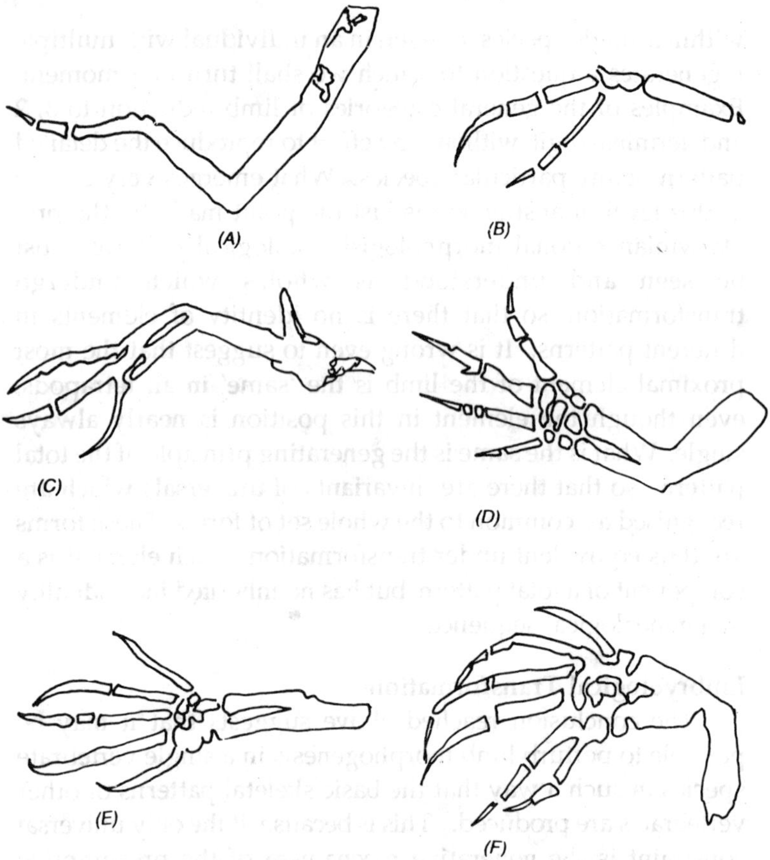

Fig. 6.7. Cartilage patterns observed in supernumerary limbs produced after 180° ipsilateral rotations of blastemas in the axolotl, *Ambystoma mexicanum (a)* one, *(b)* two, *(c)* three and *(f)* six-digits together with normal forelimb *[(d)* four-digit] and hindlimb *[(e)* five-digit] patterns.

even possible that patterns will be observed which are not found in any tetrapod, since the expectation is that the set of possible forms which the generating mechanism can produce will be greater than the set which has been realised in evolution, i.e., the potential set is greater than the historically realised set, and individual species embody this potential in their fields, independently of genotype.

The investigation of limb regeneration, in urodeles provides an opportunity for this type of observation, and we reproduce some patterns which have resulted from work with the axolotl, *Ambystoma mexicanum*, by M. Maden. A more complete study of this material is in progress, but we report we report here some observations of relevance to the present investigation. The sample provides evidence supporting the general principles developed above, illustrating how plastic is the limb field in the patterns which can be generated and reinforcing the proposition that the field is the generative unit in development and evolution.

Transformations Involving Expansion

The limb pattern an expansion of elements across the proximo-distal axis, rising to six phalanges. Such a limb is usually interpreted to result from a fusion of two four-digit limbs, with the 'loss' of a digit in the process. However, there is no need to regard this type of pattern in this way if we are using a field description of limb morphogenesis, since it is perfectly possible for the generating process to involve 'energy' levels giving a periodicity of six (or more) across the antero-posterior axis. An example of such a limb is to be found in the fossil marine reptile *Ichthyosaurus*, where the number of elements spanning the antero-posterior axis rises to six and then falls off again. The digits are not separated in this case, since the limb was of paddle type, used for swimming. It is interesting to note that no attempt is made to identify the elements of such a limb in relation to the hypothetical 'common ancestor,' except for the humerus and the radius and ulna, if the second set of elements in proximo-distal order constitute a pair radius, ulna, and pisiform if these constitute a triple.

A computer simulation of this type of limb. In this particular example, the most proximal element arises from a 'fusion' of the maxima of functions with periods 1 and 2, giving a Y-shaped chondrogenic centre. Thus the number of elements goes from one to three without the presence of a separated solution of period 2, duplicating the pattern. The type of variation seen in the bone sequence of *Ichthyosaurus* and the possibility that fusions may occur from the earliest stage of chondrogenesis further confuses the problem of identifying limb elements between species. Once again, we draw the obvious conclusion that the pattern is simply a transformation of the other limb forms conforming to the invariant generative principles of the limb field.

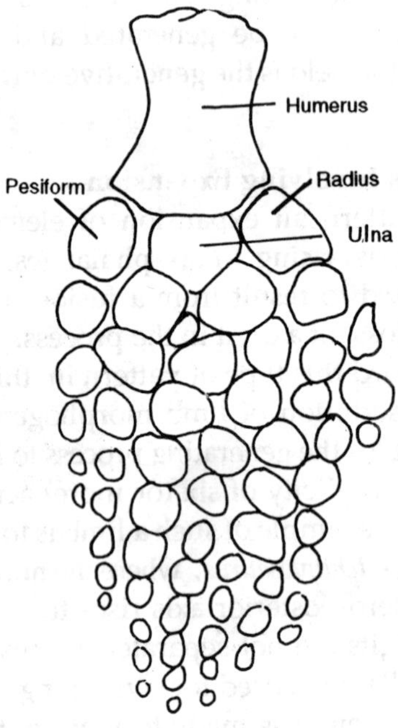

Fig. 6.8. Bone pattern in the limb of a fossil reptile, *Ichthyosaurus*.

This conclusion is also reinforced by the absence of any transient archetypal limb pattern during early limb morphogenesis, as has been demonstrated so clearly by Hinchliffe and Griffiths (this volume). However, the notion that there is inherited continuity of elements of pattern such as digits, as in Darwin's concept of continuity of descent from a common ancestor, or that there is a specific, isolable material acting as a central organising agency of biological form (Webster

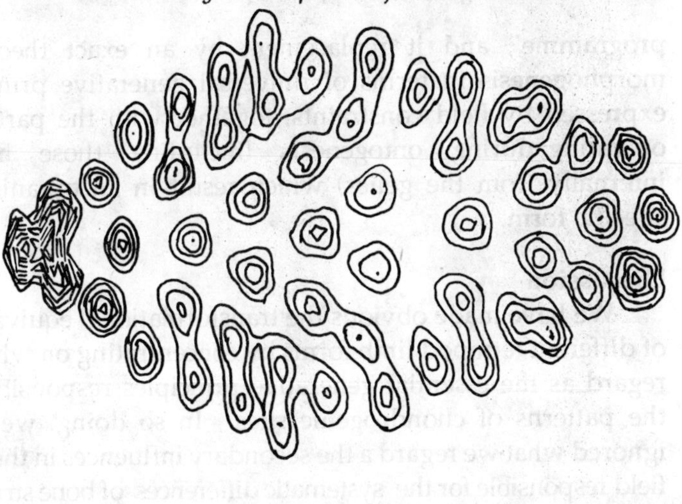

Y:	1.400											
R:	3.83	3.83	5.14	6.38	8.77	9.94	11.09	12.23	11.09	9.94	8.77	6.38
F:	0.80	0.81	0.83	0.86	0.90	0.94	0.97	1.00	1.00	0.95	0.92	0.85
A:	0.00	2.00	2.50	5.50	8.50	12.00	15.50	18.50	26.50	33.50	39.00	45.50
N:	1	1	2	3	5	6	7	8	7	6	5	3

Fig. 6.9. **Contour plot of a pattern with higher harmonic solutions (eigensolutions) than that with a periodicity of 5, simulating the pattern and showing partial contour fusions which could result in variability of bone pattern.**

and Goodwin, 1982), as in Weismann's concept of the germ plasm, dies hard. The extremely valuable studies of Nieuwkoop (this volume) show that the latter notion must also be abandoned as a general principle, since in urodeles there is no continuity of germ plasm from generation to generation, its appearance during embryogenesis being an epigenetic phenomenon. It seems likely that this will turn out to be the general case, with different species occupying different positions on the regulative-mosaic spectrum, so that the classical instances of apparent germ plasm continuity are resolvable in terms of early mosaicism in oogenesis, and earlier regulative phase nevertheless occurring. These observations all lead in the direction of an abandonment of the rather simplistic notions of the inheritance of sufficient causal elements of morphogenesis in any entity with the properties of a central organising agency, now called the 'genetic programme', and it replacement by an exact theory of morphogenesis in terms of universal generative principles expressed by field constraints, together with the particular operating during ontogenesis (including those arising internally from the genes) which result in an organism or specific form.

Discussion

We have made obvious the transformational equivalence of different tetrapod limb forms by concentrating on what we regard as the essential generative principles responsible for the patterns of chondrogenic sites. In so doing, we have ignored what we regard a the secondary influences in the limb field responsible for the systematic differences of bone structure observed across the antero-posterior axis, and the factors responsible for the differential growth of the chondrogenic sites in the proximo-distal axis. That these are considered as secondary, not primary, is one feature that distinguishes our treatment of limb morphogenesis sharply from a model such as that of Wolpert *et al.* (1975), in which the limb elements are derived from primary axial gradients by a process of 'interpretation' and there is no prepattern of elements in a field isomorphic with the manifest limb pattern. A major difference between these views is in the number of parameters involved in generating the pattern and in changing from one pattern to another. Applying Wolpert's chick limb model to the pentadactyl limb, and using the threshold mechanism of Lewis *et al.* (1977), about 24 parameters are needed to generate 5, discrete element across the antero-posterior axis. Each digit gained' or 'lost' involves changes in several parameters, for the threshold points must be properly positioned relative to the morphogen gradient to produce discrete, separate chrondrogenic sites. The proximo-distal pattern requires more threshold points in another monotonic gradient involving some 35 parameters or more. Clearly, genetic programmes of this type are expensive on particulars because there are no generative universals; and since, in such models, parameters are specified by genes, we would expected a large number of mutants affecting the details of limb morphology, for which there is little evidence. A polydactylous mutant with six digit requires either the introduction of a new threshold or a change in the gradient such that a 'hidden' threshold is revealed, in which case it might be concluded that the 'common ancestor' had six digits.

On the whole, the evidence' seems strongly to favour a more robust pattern-forming mechanism involving fewer parameters, whose basic characteristic is the generation of spatially ordered arrays (whole patterns) of repeating elements which are primarily the same but differ by secondary variations superimposed on the primary pattern generator. This takes us back to the prepattern fields of Stern (1954) and Turing (1952), and away from positional information and interpretation, whose specific objective was to replace prepatterns by monotonic gradients and thresholds wherever there is evidence that the units produced are different from one another. The model proposed in this paper, involving a field theory, which generates a morphogenetic prepattern of a type different from those based upon reaction and diffusion but with some fundamental characteristics in common, is economical on parameters, since the basic pattern varies in response to the proximo-distal *'energy'* profile. This profile can be described by curves of three parameters, only one of which needs to be varied to generate the range of patterns shown. Genes are assumed to be involved in specifying the details of these curves, but the rest of the pattern-forming mechanism is assumed to be universal to the tetrapods. We must then assume a secondary mechanism cross the antero-posterior axis to introduce variations in the intensities of the order parameter, u, leading to asymmetries in the units across this axis.

Tetrapod limbs, as a general category of vertebrate structure, are characterised in such an analysis by a high-level constraint relating to the generative particulars which distinguish limbs from fins and other appendages, presumably involving genes implicated in processes such as limb bud formation and chondrogenesis. However, as well as these particulars, the basis of evolutionary change, there will continue to be universal mechanisms involved which, in our model, are the morphogenetic fields themselves, with solutions characterised as eigensolutions in our formulation. Thus, at every level of a generative analysis, there is a combination of universal and particular factors at work.

The view of biological form advanced here is based upon the general scientific principle that diversity of structure is logically intelligible only if there is some unity or invariance which limits not only observed diversity, but possible or potential diversity. In the case of the typical vertebrate form which has come to be known as the pentadactyl limb, this invariance is to be found in the generative principles which are involved in primary limb morphogenesis. Our particular model involves constraints deriving primarily from the mechanism whereby a continuous 'energy' profile selects a series of eigen-function solutions of a field arising in a distally propagating domain of competence in the developing limb bud. In consequence, limb forms are defined by a series of discrete chondrogenic sites whose antero-posterior arrays differ in number of elements by a small integer (usually one) as one moves distally. If limbs commonly occurred with elements increasing distally in a sequence such as 1, 2, 4, 8, 16, 32, ... then different generative rules would be involved. Such a sequence describes the number of cells in holoblastic cleavage, which has been shown to be describable by a field treatment similar to that used here for vertebrate limb morphogenesis, but involving different selection (generative) rules on the field solutions. Thus, we see that the general field approach to biological form gives a hierarchy of constraints within a global theory whereby different forms can be distinguished from one another while till united by general principles of spatial organisation as embodied in the field treatment. In this way both the unity and the diversity of biological forms may be grasped, and a generative basis provided for a rational taxonomy, as envisaged by the pre-Darwinian rational morphologists.

Another constraint arising from our field description of limb morphogenesis is very similar to that which Holder has deduced in relation to the polar co-ordinate model of limb formation. His prediction is that limbs will not be found whose primary patterns have external digits larger than internal ones. Anomalies such as the lobster-claw syndrome are excluded because they arise from a secondary process which effectively destroys part of the primary pattern. A similar prediction can be made for the pattern-generating mechanism which we describe in this paper. This is that tetrapod limbs will not be found with more phalanges in the external digits than in the internal one. The reason for this is that there cannot be an eigensolution which has blank spaces in the middle of an antero-posterior array. The converse can certainly occur, as illustrated, where the interior three digits each have one more phalange than the external digits, because of a drop in the 'energy' profile distally. With antero-posterior asymmetry, there could be a very long external digit on one side of the limb, as in certain species of whale, but not on both.

The approach to morphogenesis described in this paper has the further consequence that the historical sequence of form emerging during evolution is logically secondary to an understanding of the generative principles defining the potential et of these forms and their transformations; so that Darwinian taxonomy, which is genealogy, is secondary to rational taxonomy, and both are logically derivative from ontogenetic principles. The external selection principle on historically realised forms which is added by the functional or stability considerations embodied in the concept of fitness, while of obvious importance in many instances, is supplementary to those which operate within organisms in the generative process itself, and certainly can never be used as a causal substitute for the latter. Within the rationalist approach to biological morphology being proposed, the interesting possibility emerges that a stability constraint (such as 'fitness') on the set of possible morphological transformations may embody a general principle of directed or time-dependent transformation of the type which in physics gives such directional stability criteria as the second law of thermodynamics, imposing temporal asymmetry on the time-reversal symmetries of mechanics. It is evident that a much more highly elaborated theory of form and stability than that current available is required before this type of theory can emerge, although it remains a part of the research programme which follows from the principles underlying the treatment of biological form and organisation advanced in this paper.

7

Evolutionary Significance of Developing Limb

The pattern of the vertebrate pentadactyl limb which is such a striking feature of the skeleton of tetrapods constitutes *the* classic example of homology. As early as 1849, in his book *On the nature of limbs*, Richard Owen drew attention to the fundamental similarity of pattern in tetrapods, regardless of the use to which the limb was put, whether for flight, for swimming or for running. As an opponent of evolution, Owen had no explanation for this fundamental similarity — which he termed 'homology' — but to Darwin the pentadactyl pattern provided a splendid illustration of the principle of adaptive radiation. What can be more curious, asks Darwin (1872), 'than that the hand of man formed for grasping, that of a mole, for digging, the leg of a horse, the paddle of a porpoise and the wing of a bat, should all be constructed on the same pattern and should include similar bones and in the same relative positions'. To Darwin, the homology of the vertebrate limb, regardless of its various functional adaptations, indicated descent from a common ancestor.

In due course, as discoveries were made in fossils of the structure of the primitive amphibian limb skeleton (e.g., in *Eryops*) and of the Crossopterygian fish fin paddles Gregory and his colleagues were able to put forward a plausible scheme of evolutionary transmission of fish fin into tetrapod limb. The homology between fin paddle and limb is more evident at the proximal end of the limb where amphibian stylopod (humerus, or femur) and zeugopod (radius, ulna or tibia, fibula) elements are clearly represented in the Crossopterygian fin paddle skeleton. More distally, from the wrist or ankle outwards, the correspondence has been more debatable, with little agreement between the different authorities as to which elements of the fish paddle survive, how they may be divided up, or whether the tetrapod digits represent newly evolved ('neomorph') structures.

Abbreviations for all Figures

Be, basal commune; C, central carpals or tarsals; F, fibula; f, fibulare; fc, falciform; Fe, femur; H, humerus; i, intermedium; p, pisiform; R, rad- ius; r, radiale; T, tibia; t, tibiate; U, ulna; u, ulnare: Y, element; Y, according to Holmgren (1933), restricted to urodeles; 1-5, distal tarsals (dt) and carpals (dc);

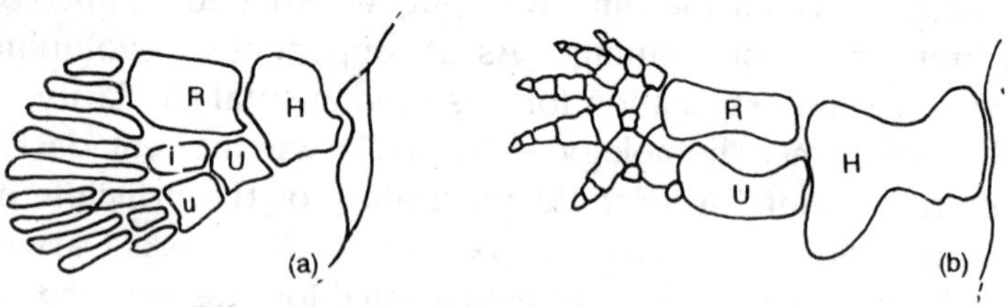

Fig. 7.1. Gregory's theory (Gregory & Raven, 1941) of the transformation of a Crossopterygian (*Sauripterus*) pectoral fin paddle skeleton *(a)* into the pentadactyl limb *(b)* of the primitive amphibian.

A number of attempts have been made to construct a fundamental pattern — or archetype — for the pentadactyl limb based on the skeletal pattern in the primitive amphibians. The so-called canonical elements of the carpus according to Holmgren consist of four proximal elements, radiale, intermedium, ulnare, pisiform), four centra elements and five distal carpals, one at the base of each of the five digits. The precise composition of the pattern which makes up the pentadactyl limb is difficult to define, since digital number and phalangeal formula are very variable between different species, as in specialised forms such as adult bird wing and horse foot. There has been a long held (and pre-Darwinian) assumption that in development the general precedes the specialised in the formation of structure, and it is still widely believed that, however specialised the adult limb may be, it passes through an archetypal pattern in its development. Holmgren thus considered that the canonical elements of the carpus could be identified in limb development throughout the tetrapods, with the exception of the urodeles, for reasons discussed later. In his wide-ranging surveys of many species of anurans, reptiles, birds and mammals, the same archetype is described in the prechondrogenic pattern of the limb bud.

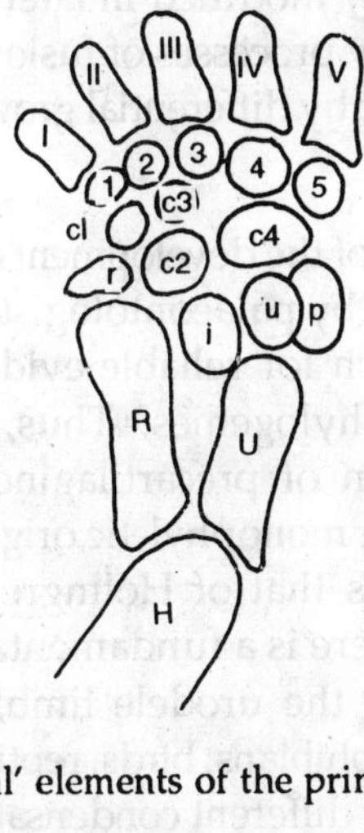

Fig. 7.2. 'Canonical' elements of the primitive archetypal carpus.

This approach is stated, in a general way, by Darwin, who wrote that 'The forelimbs, which once served as legs to a remote progenitor, may have become, through a long course of modification, adapted in one descendant to act as hands, in another as paddles, in another as wings: but...the forelimbs will not have been much modified in the embryo of these several forms; although in each form the forelimb will differ greatly in the adult stage'. Essentially this view is still echoed today in Jarvik (1980), who states that ....ontogenetic development of the extremities proceeds from a given number of metameric elements. According to a common opinion originating from Gegenbaur the extremity skeleton in tetrapods is constructed from a definite number of elements which are arranged, in the hand (manus) as well as the foot (pes) according to an original common ground plan'. Normally the limb skeleton in its development goes through a process in which a prechondrogenic condensation of closely packed cells forms in undifferentiated mesenchyme, the condensation gradually chondrifying through the secretion of cartilage matrix, and thus forming a cartilage 'model', which usually later hypertrophies and is replaced by bone. According to the archetype theory, the precartilage condensation pattern represents the pentadactyl limb archetype, which becomes gradually modified in later development into more specialised forms by processes of fusion or disappearance of individual elements, or by differential growth rate of different elements.

Descriptive accounts of the development of the limb skeleton have thus been seized on by palaeontologists and comparative anatomists in their search for reliable evidence of homology on which to base their phylogenies. Thus, identification of a single archetypal pattern of precartilaginous condensations will support the view of a monophyletic origin of the tetrapods. Another interpretation is that of Holmgren (1933-1939) and Jarvik who argue that there is a fundamental difference in the condensation pattern in the urodele limb, from that in the 'eutetrapods' (anuran amphibians, birds, reptiles and mammals), which possess a distinctly different condensation pattern. These two authors argue that urodeles originate independently from the eutetrapods, and that the tetrapods are therefore diphyletic in origin. Jarvik (1980) sees the same fundamental seven-rayed pattern in the Crossopterygian *Eusthenopteron*, in the primitive amphibian limb, and in the prechondrogenic and chondrogenic pattern in the eutetrapod limb bud, for example in both the human embryonic hand and that of the lizard *Lacerta* Urodeles he regards as having a distinctly different pattern, with digits 1 and II being carried by a single basal carpal element, the intermedium ray thus being branched in urodeles, but carrying only a single digit in the eutetrapods.

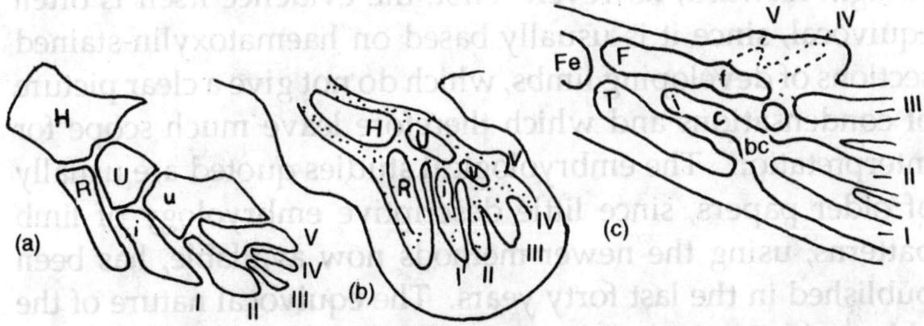

Fig. 7.3. According to Jarvik (1980) the fin paddle skeletal pattern of the Crossopterygian *Eusthenopteron (a)* is repeated in limb development in the pre-and chondrogenic pattern of the 'eutetrapod' group (anurans, reptiles, birds and mammals) as represented by the lizard, *Lacerta (b)*, Jarvik claims the pattern of the intermedium ray and digits I and II is fundamentally different in urodeles, as represented by *Triturus*.

Again, descriptive embryology has been used in the recent discussion as to the origin of birds and the significance of *Archaeopteryx*. Ostrom (1976) has claimed there is strict homology between the three digits of certain coelurosaurian dinosaurs and *Archaeopteryx* as part of his argument for a dinosaur origin for birds. These three digits are clearly anterior digits (I-III) in dinosaurs. The question then arises as to whether the three surviving digits in modern birds are anterior as would be shown if there were embryological evidence of the existence of two vestigial digits (IV, V) posteriorly. In this case, the embryological evidence provided by the digits is not conclusive, since only one posterior vestigial digit can be found. A different viewpoint on bird evolution is provided by Walker (1972), who claims that birds and crocodiles are closely related, partly on the basis of the similarity of the developing pattern in the carpus and tarsus. This similarity is not however, supported by more recent evidence of birds carpus and tarsus development. These three examples illustrate how the descriptive embryology of the limb may be used in the attempt at construction of phylogenetic relationships based on homology.

The use of embryological evidence in such a way is not straight-forward, however. First, the evidence itself is often equivocal, since it is usually based on haematoxylin-stained sections of developing limbs, which do not give a clear picture of condensations and which therefore leave much scope for interpretation. The embryological studies quoted are usually of older papers, since little descriptive embryology of limb patterns, using the newer methods now available, has been published in the last forty years. The equivocal nature of the older evidence is apparent in reviewing the literature on the development of the pattern in the bird wing. Here, the number of precartilaginous carpal condensations is differently identified by different authors. Estimates range from 13 to 4. Since the number of cartilaginous elements in the wrist is small, authors such as Montagna and Holmgren invoke a process of fusion and regression to explain the reduction from the larger number of condensations. However, a recent investigation of the carpal condensation pattern, using autoradiography, revealed only four condensations, corresponding to the number of cartilaginous elements, and there appeared to be no process of fusion and regression. These findings appeared to question the existence of an archetypal condensation pattern in the wrist in birds, and thus a wider survey of the condensation pattern was attempted using autoradiography in a number of representative species: the axolotl (urodele), *Xenopus* (anuran) and the mouse. The previous findings in the chick limb are first briefly reviewed; the new findings are presented and discussed in relation to the hypothesis of a limb archetype, and to control mechanisms of the development of limb skeletal pattern.

Autoradiographic for the Analysis of the Prechondrogenic Pattern

The condensations are ill-defined using standard histological methods, which also fail to distinguish prospective chondrogenic from tendon and myogenic condensations. A more specific method of identifying chondrogenic condensations is clearly needed. Using $33SO_4$ the chondroitin sulphate component of cartilage intercellular material can be labelled. Searls (1965a) showed that if chick limb buds are labelled with $35SO_4$ and processed for light microscopy, all the radioactivity left in the limb buds is bound to proteoglycan, mainly chondroitin-4 and chondroitin-6 sulphates. While some of this is initially localised in prospective soft-tissue and myogenic regions, chondroitin sulphate synthesis eventually falls off in the non-chondrogenic regions and increases in the prechondrogenic condensations. The prechondrogenic pattern representing the future zeugopod elements can be detected by autoradiography at late stage 22, which is slightly before the increased packing of the condensations can be identified by histology. By application of 80 μCi $35SO_4$ to the chick embryo (4-7 days) vitelline circulation for 4h the developing pattern of wing and leg prechondrogenic condensations and chondrogenic elements was mapped out. The resulting autoradiographs produced a more specific and clearly defined map of the condensations and their transition into cartilage elements than in the classical histological accounts.

We attempted to apply similar methods to *Xenopus* and axolotl *(Ambystoma)* larva, but without initial success. Direct injection proved too difficult, while larvae immersed in various concentrations of $35SO_4$ in freshwater failed to show much radioactivity in their limbs. Finally, we cut off the developing limbs and cultured them for 12 h in Holtfreter's solution to which $35SO_4$ was added at an activity of 80μ Ci per ml. Details of fixation, use of stripping film and exposure were the same as for the chick limb buds. The results were more variable and the condensations less clearly defined than in the chick limb buds and there tended to be more background activity in the soft-tissue regions. Nonetheless, the patterns were more precise than those obtained by histology, though in some instances the number and form of elements is clearly debateable.

In the case of the mouse, pregnant mice of 13, 14 and 15 days were injected intraperitoneally for 12 h with 500μ Ci of $35SO_4$, after which the embryos were obtained, and the limb buds then treated in the same way as chick limb buds. The day on which the copulation plug was found was designated day 1. The technical work was carried out by Dr. M. Hicks. A similar study was made by Milaire (1978) using the $35SO_4$ method, which he found to produce a pattern of chondrogenesis slightly in advance of histochemical methods involving staining for proteoglycans. Where elements correspond they are labelled according to Holmgren's interpretation, but this does not mean that Holmgren's conclusions about homology are accepted.

Prechondrogenic Pattern in the Chick Wing and Leg Bud

This pattern has already been described and here only the main features will be reviewed. Autoradiography shows carpal condensations starting to appear at stage 27 (5 days) at the same time as the condensations of metacarpals 3-5. Four carpal elements are clearly defined by stage 28 (6 days); radiale, ulnare, pisiform and a single distal carpal at the base of metacarpal 3. After stage 30, the ulnare (by now chondrogenic) ceases to be active synthetically, and soon becomes pycnotic and disappears. It is the only chondrogenic element to disappear. Only condensations which later become chondrogenic are present, and no condensation fusion can be seen. Autoradiography does not reveal the 'missing' digit in the chick: the three main digits (conventionally termed II-IV after Montagna) and a small posterior metacarpal (V) are present, but there is no sign of a fifth digit.

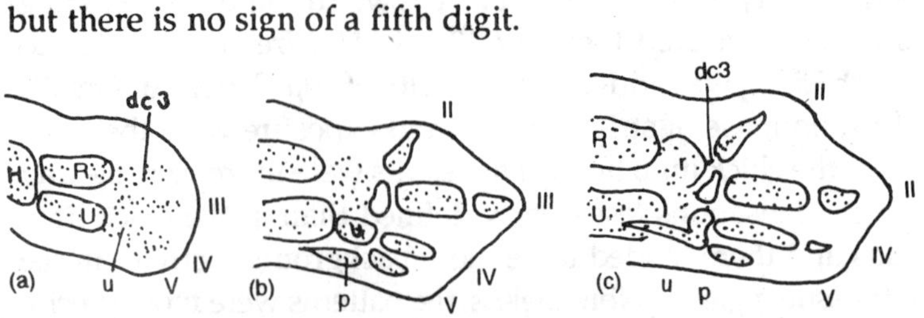

Fig. 7.4. Development of the pattern of prechondrogenic condensations and cartilaginous elements in the chick forelimb. a Stage 27 (5 days); *(b)* stage 28 (5.5-6 days), and *(c)* stage 30 *(7 days).*

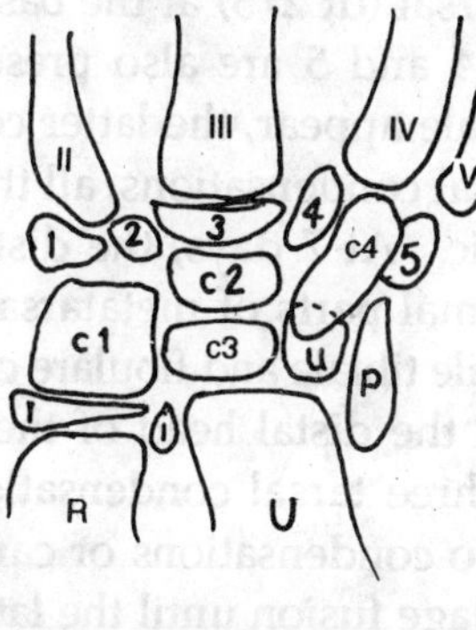

Fig. 7.5. Montagna's (1945) interpretation of the prechondrogenic pattern of the chick carpus at 6 days.

This account is very different from the classical description which is summarised in such standard accounts of avian embryology. All the previous accounts, though differing from each other in points of detail, describe a pattern of carpal condensations reasonably close to the ancestral amphibian carpus. Thus, at 6 days Montagna describes four proximal elements (radiale, intermedium, ulnare, pisiform), the four central elements (C_1-C_4) and finally five distal carpals (dc_1-dc_5). A process of fusion of some condensations and non-chondrification of others then accounts for the pattern of chondrogenic elements. The 13 condensations thus become 4 cartilages. In Holmgren's account, 11 condensations are described, the identification of central carpals differing from that of Montagna. According to other accounts the number of condensations is considerably greater than the number of carpus cartilages.

In our view, the condensation number is the same as the cartilage number, and there is no process of condensation fusion or of non-chondrification. The carpal condensation pattern is specialised and quite different from the archetype of primitive amphibians.

In the leg bud, autoradiography shows that prechondrogenic

tarsal elements elements first appear at stage 26 with the fibulare and a single distal tarsal (dt 2/3) at the base of metatarsals 2 and 3. Metatarsals 4 and 5 are also present. At stage 27, metatarsal 1 and tibiale appear, the latter completing the full complement of 3 tarsal condensations, all three of which then become chondrogenic. At 7 days, the distal tarsal cartilage fuses with the proximal parts of metatarsals 2-5 to form the tarso-metatarsus, while tibiale and fibulare cartilages fuse with each other and with the distal head of the tibia to form the tibio-tarsus. Only three tarsal condensations appear, all of which chondrify. No condensations or cartilages disappear, and there is no cartilage fusion until the later (7-day) stage.

Once again, this account is very different from classical descriptions. Lutz (1942) claimed to find ten condensations, including a complete row of five distal tarsals. Holmgren (1955) also identified ten elements, though he regarded several as being already fused into two large condensations which correspond with our dt 2/3 and our tibiale. Both these classical accounts describe a pattern close to the ancestral amphibian carpus. In our view, the more specific method of autoradiography shows that the number of tarsal condensations is much smaller, and as in the wing, the tarsal pattern is already specialised and quite different from the archetype.

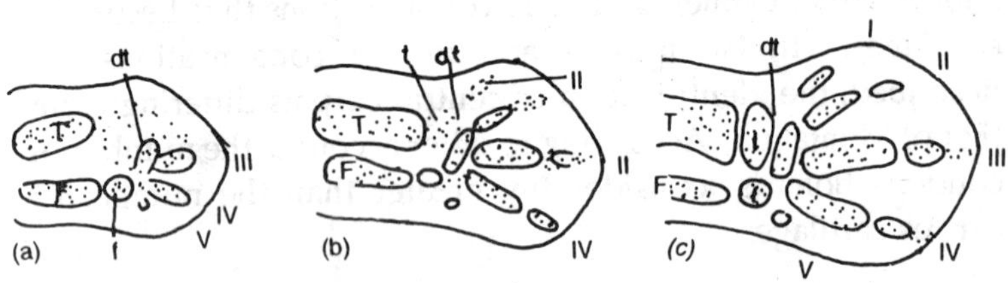

Fig. 7.6. Development of the pattern of prechondrogenic condensations and cartilaginous elements in the chick hindlimb. *(a)* Stage 26 (5 days): *(b)* stage 27; *(c)* stage 28. Drawn from autoradiographs of the pattern of $33O_4$ incorporation.

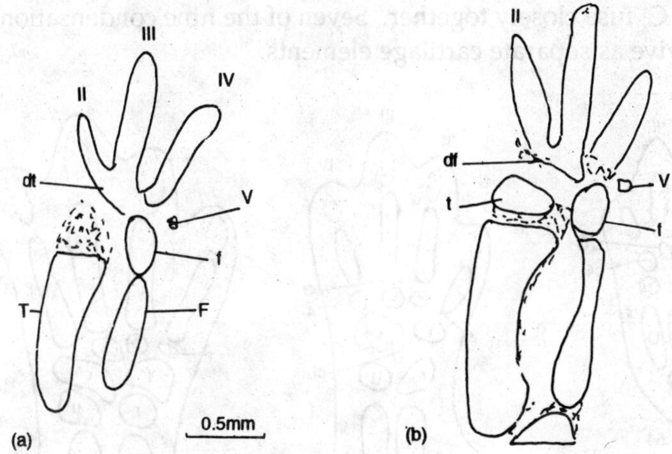

Fig. 7.7. Autoradiographs of 33SO_4 uptake by the chick hindlimb, *(a)* stage 26/27; *(b)* stage 27/28.

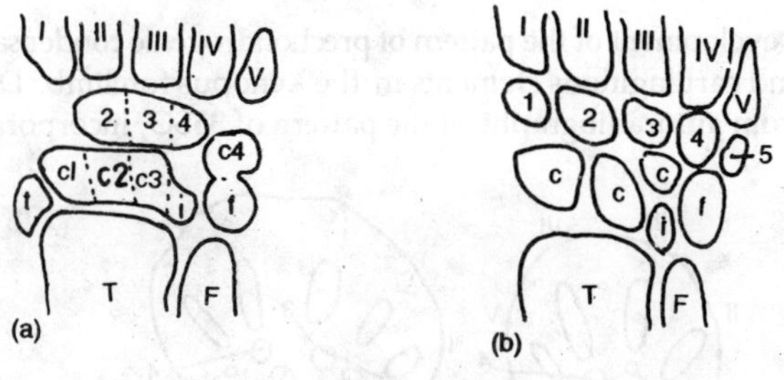

Fig. 7.8. Interpretations of the prechondrogenic pattern of the avian tarsus by Holmgren *(a)* 1955 - chick) and by Lutz *(b)*, 1942 - *Anser)*.

The Prechondrogenic and Chondrogenic Pattern in Xenopus

There appear to be nine condensations in the *Xenopus carpus*, though they do not appear simultaneously. The two proximal elements are radiale and ulnare, while distally, each of the four metacarpals has an associated distal carpal, though 3 and 4 fuse with their adjacent metacarpals. There are three central elements, all of which survive as cartilages, though C_1 and C_4 fuse closely together. Seven of the nine condensations survive as separate cartilage elements.

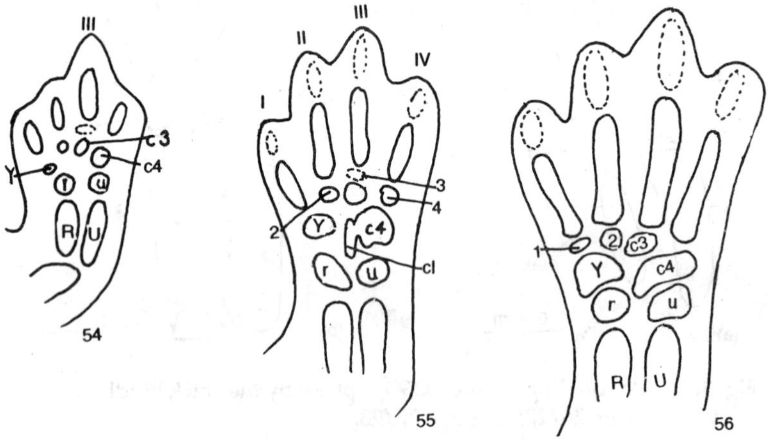

Fig. 7.9. Development of the pattern of prechondrogenic condensations and cartilaginous elements in the Xenopus forelimb. Drawn from autoradiographs of the pattern of $35SO_4$ incorporation.

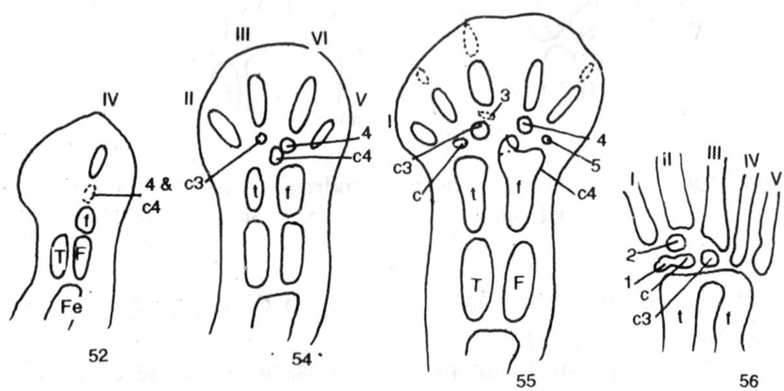

Fig. 7.10. Development of the pattern of prechondrogenic condensations and cartilaginous elements in the *Xenopus* hindlimb. Drawn from autoradiographs of the pattern of 33SO4 incorporation.

This interpretation is reasonably similar to that of Holmgren who identifies all these nine condensations, but finds in addition an intermedium and a centrale 2, which we did not find. The condensation pattern we found closely resembles the cartilage element pattern, and not the archetypal pattern in primitive amphibians, since there are too few elements and the pattern in the proximal part of the carpus is quite different.

In the hindlimb, a description of the pattern is complicated by the appearance of the 10 tarsal condensations at different times. The two proximal elements are the elongated tibiale and fibulare. Distally, each of the five metatarsals has an associated distal tarsal, and these appear in the order 4, 3, 2 and 5, 1. Of these, 3, 4 and 5 appear to fuse with their metatarsals, without becoming a separate chondrogenic element. There are three central elements, one each at the base of digits II and III, and a third which fuses with the distal end of the fibulare. Thus only six of the ten condensations form separate chondrogenic elements.

Holmgren's reconstruction is comparable to our stage 54 with the exception of his distal 3, which appears later, at stage 55. Though interpretation is not easy on account of condensation and cartilage element fusion, the carpus pattern does not homologise readily with the archetype. There are too few elements, both proximally, where the intermedium is absent, and also centrally, where only three of the four archetype centrals can be found.

The Prechondrogenic and Chondrogenic Pattern in the Mouse Forelimb

Descriptive accounts of the developing chondrogenic pattern in the mouse have been published by a number of authors. The chondrogenic elements of the carpus are clear by 14 days, when the pattern is essentially that of the adult. If the falciform is excluded, there is a total eight elements, three

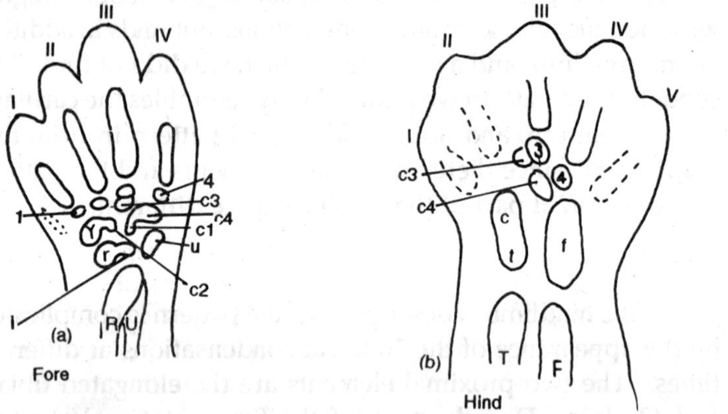

Fig. 7.11. Holmgren's (1933) interpretation of the prechondrogenic pattern of *Xenopus* carpus *(a)* and tarsus *(b)*. This interpretation conforms with the archetype.

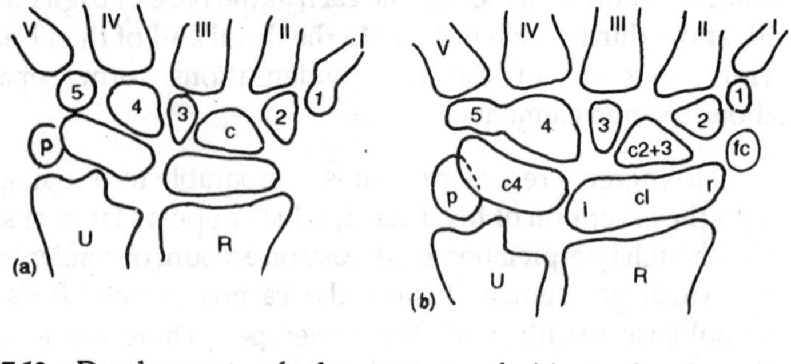

Fig. 7.12. Development of the pattern of *(a)* prechondrogenic condensations (13 days) and *(b)* chondrogenic elements (14 days) in the mouse forelimb. Drawn from autoradiographs of $33SO_4$ incorporation partly from Milaire (1978) and partly from autoradiographs obtained by M. Hicks. The labelling in *(b)* follow Holmgren's (1952) interpretation which identifies two elements (C_2 and 3; 1, C1, r) as formed by fusions involving central condensations, thus enabling the pattern to conform with the archetype (see text).

proximal (including pisiform), one central and four distal carpals, one for each digit except for IV and V which have a common distal carpal. Autoradiography reveals that, at 13 days, each of these chondrogenic elements is preceded by a condensation, except that at this stage there are separate distal carpals 4 and 5.

Holmgren identifies all these condensations, but in addition claims two chondrogenic elements are formed by fusion. These are (i) the element distal to the radius, claimed to be a fusion of radiale, central 1 and intermedium, and (ii) the central element, claimed to be a fusion of centrals 2 and 3. According to our autoradiographs, both these are single condensations. A careful study by Milaire (1978), also using $35SO_4$ considers the first as a single, but the second as a double condensation (though his 12-day autoradiographs suggest continuity between C_2 and C_3). Dalgleish (1964), in a purely histological study of the mouse limb, also identifies all the eight carpal chondrogenic elements as single condensations at 12 $^1/_2$ days, the falciform appearing at 13 days. Dalgleish concludes that' ... no real evidence is found for the previously reported four centrals other than the single definitive centrale of the mouse hand. Several structures... have been mistakenly considered such vestiges. Lack of such evidence casts doubt upon the conclusions of others that, phylogenetically, the mouse limbs are amongst the oldest tetrapod appendages.'

The adult mouse carpus is relatively unspecialised and its form is very similar to the standard pattern in mammals with unspecialised limbs and in primitive reptiles. Even so, the condensation pattern of the mouse carpus more closely resembles that of the resulting cartilages than it does the archetypal pattern represented by the primitive amphibian carpus. To give one example: at eight, the number of mouse carpal condensations is too low.

The Prechondrogenic and Chondrogenic Pattern in the Axolotl Forelimb and Hindlimb

In the stage 43 forelimb, radius and ulna and digits I and II are clearly defined. In the carpus, only the basal commune (bc) is well-defined, but there are three areas of increased

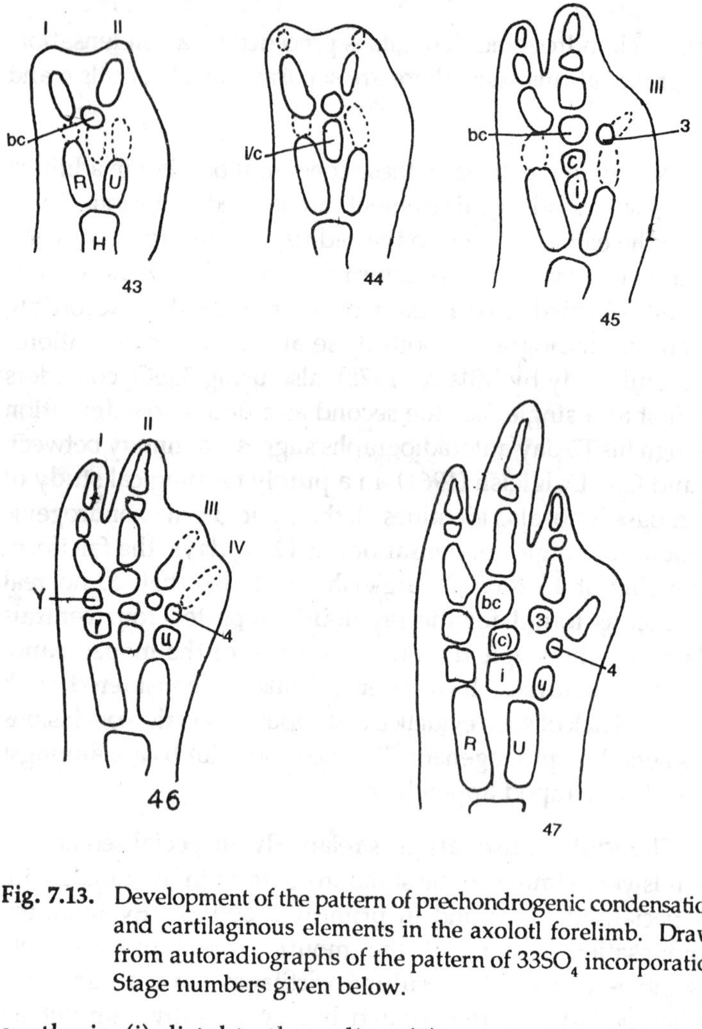

Fig. 7.13. Development of the pattern of prechondrogenic condensations and cartilaginous elements in the axolotl forelimb. Drawn from autoradiographs of the pattern of 33SO$_4$ incorporation. Stage numbers given below.

synthesis, (i) distal to the radius, (ii) proximal to the bc, (iii) distal to the ulna. Gradually these three areas resolve themselves as (i) radiale and 'Y' by stage 46, (ii) intermedium and central, by stage 45 and (iii) ulnare (stage 46), distal carpal 3 (stage 45) and 4 (stage 46). The additional digits III and IV become clear by stage 46. The metacarpai of digit I develops a distinct articulation towards bc by stage 44. By stage 47, the chondrogenic pattern of the *larval limb* (essentially that of the adult also) has been completed. Autoradiography does not clearly indicate that digit I is part of a single ray including the radius, on the other hand it does not support the view that digits I and II and the intermedium ray form initially as a single Y-shaped bifurcation, with the b.c. at the base of the digits. The head on the metacarpal 1 which articulates with b.c. is secondary, a conclusion which Thomson (1968) also reaches from the descriptions of other urodeles by Sewertzoff (1908) on *Triturus* and Schmalhausen (1910) on *Hynobius* (='Salamandrella'). The condensation pattern in the axolotl is very similar to the final chondrogenic pattern (although the intermedium and central separate from a single initial condensation), and there is no fusion or loss elements.

The hindlimb in the axolotl follows an almost exactly similar pattern of chondrogenesis to the forelimb. The eight carpal elements all have precise tarsal equivalents, although in addition there is a fifth digit which has an associated distal tarsal element.

The developing chondrogenic pattern in urodeles has been used by Holmgren and Jarvik as important evidence in their separation from the 'eutetrapod' group. Jarvik (1964) states that in urodeles 'the intermedium ray is branched and carries two digits (I & II) in urodeles, whereas it . . . is unbranched and carries one digit (I) in other tetrapods. .. There is always a basal commune in the urodelian limb. The autoradiographic pattern does not support this view, and thus provides no ontogenetic evidence for the phylogenetic separation of urodeles from eutetrapods. But since the urodele pattern is very different from that of the species chosen as representative of anurans (which lack an intermedium), birds and mammals; and since each of these patterns is specialised, prechondrogenic pattern cannot be used as evidence of homology in deciding the possible monophyletic origin of tetrapods.

Possible Mechanisms of Pattern Change in Evolution.

The archetypal condensation pattern theory must assume

as a logical consequence that the specialisation of the tetrapod limb occurring during its adaptive radiation involves developmental mechanisms which affect only the postcondensation processes, such as growth and chondrification. Thus, if evolution favours, as it frequently does, a large tibia in relation to the fibula, then there must be differential growth during chondrogenesis of tibia relative to fibula from initially rather similar-sized condensations. Alternatively elements may fuse or disappear, as happens with several chondrogenic elements in *Xenopus* carpus and tarsus and the ulnare in the chick wing. These processes seem to be the consequence of a growth programme — presumably genetically controlled— inherent in the condensation. If isolated, and grown by culture *in vitro*, such elements grow to their characteristic size, whether large or small. Thus Hicks (1982 isolated early chondrogenic rudiments of chick tibia and fibula and found that, when grown *in vitro*, the growth rate of the tibia was much greater than the fibula.

While differential growth of cartilage elements clearly exists, it is quite clear that genes control, directly or indirectly, the initial pattern of condensations. There are thus no good grounds for making the archetype in some way sacrosanct from modification by natural selection acting on genetically controlled variation. It is only in a few 'laboratory' species such as the mouse that we know much about the effects of gene mutations on the process of skeletogenesis, but it is clear from the studies of Gruneberg (1963) that the initial condensation pattern is susceptible to modification by gene action. In *Oligosyndactylism* the footplate is narrower than normal, so that the digital condensations are either squashed closer together, or digit II may be missing completely. In the luxate group, some limb bud mesenchyme is in effect frequently shifted from making one proximal condensation to making another distally. Thus in the hindlimb bud, the tibial condensation is much reduced in size, while in the footplate there is an anterior excess of mesenchyme which goes to form an enlarged anterior

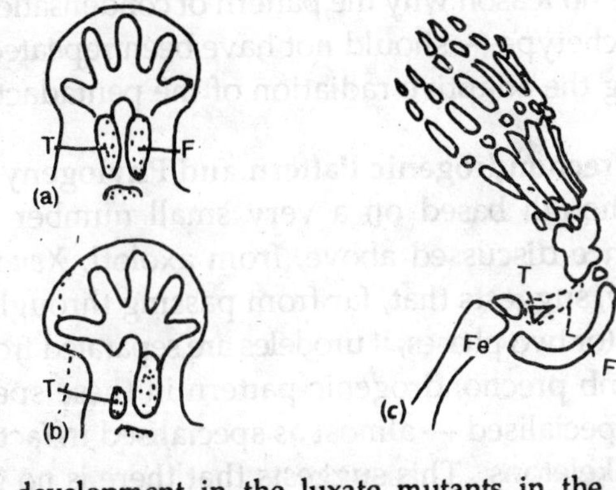

Fig. 7.14. Limb development in the luxate mutants in the mouse. Mesenchyme is effectively shifted from making tibia to making enlarged or additional preaxial digits (a) Normal and (b) luxate leg bud at 12 $^1/_2$ days. A typical resultant skeleton, in this case Strong's luxoid, is illustrated in (c), in which the tibia is reduced and there is an additional preaxial digit. L = ligament.

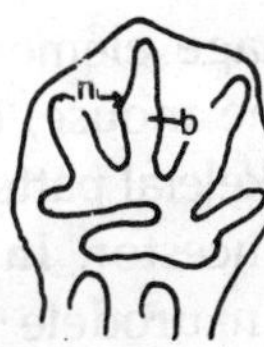

Fig. 7.15. Limb development in the brachypod mutant in the mouse. Mesenchyme is misallocated between metacarpal and phalangeal condensations. Normal *(n)* and brachypod (b) positions of metacarpalphalangeal future joint in 13-day leg bud. The brachypod metacarpals are abnormally small.

(first) digit and frequently also an additional preaxial digit. One of the effects of the *brachypod* mutation is to misallocate mesenchyme between digital condensations: the mutant metatarsal condensation is smaller than normal, the additional mesenchyme being allocated to the adjacent phalangeal condensation. These few examples cited from a mass of available evidence make it clear that there exists genetically controlled variation of the normal condensation pattern. There is thus no reason why the pattern of condensations — including the archetype — should not have been repeatedly remodelled during the adaptive radiation of the pentadactyl limb.

The Prechondrogenic Pattern and Phylogeny

Though based on a very small number of species, the evidence discussed above, from axolotl, *Xenopus*, chick and mouse, suggests that, far from passing through an archetypal phase (or two phases, if urodeles are separated from eutetrapods) the limb prechondrogenic pattern in these species is already very specialised — almost as specialised in fact are their adult limb skeletons. This suggests that there is no general pattern of condensation, common to all tetrapod groups. In consequence, there is no justification for constructing a eutetrapod group on the basis of similarity of condensation pattern in *Xenopus*, chick and mouse. From the avian evidence it is very clear that the pattern of condensation in ankle and wrist regions is much more closely related to the later avian skeletal pattern of cartilage elements than to condensation pattern in *Xenopus*, or the mouse, or for that matter, to the adult limb or fin paddle skeletal pattern in primitive amphibian or Crossopterygian fish ancestor. In view of these differences, the argument whether in urodele limb ontogeny from the condensation stage digits I and II have a common base on the basal commune or whether this is secondary ceases to have important phylogenetic implications, since there is no homologous carpus or tarsus pattern throughout the eutetrapod group from which the urodele pattern can be distinguished. This conclusion rules out the condensation pattern as useful evidence in constructing overall schemes of tetrapod evolution though it may still provide evidence of homology in relation to particular problems within the tetrapods, such as the relationships of birds and crocodiles. In our view, the identification of a condensation archetype has arisen because investigators have super-imposed their preconceptions on histological evidence which was unclear.

As attention currently focuses on models which generate limb patterns, this may seem rather a negative conclusion. But such models must in the end aim to generate actual limb patterns, and it is perhaps worth emphasising the rather specialised and complicated way in which the final pattern of tarsus and carpus emerges. At present most of the models simply omit carpus and tarsus. Finally, it is worth pointing to evidence in the mouse of genetic modification of condensation pattern, which provides a possible mechanism for the modification in evolution of limb skeletal structure through selection pressure.

8

Regeneration in Amphibian Limbs

Many organism or parts of organisms are capable of pattern regulation not only during embryonic stage, but throughout life. The domain within which regulation can occur has been designated a 'field' and it has been recognised for many years that field can be roughly categorised according to their mode of regulation following surgical intervention. One category contain those systems which display morphallactic regulation, where a part of the pattern can reorganise to form a miniature but complete pattern. The other category consists of those systems showing epimorphic regulation, where a part of the pattern can add on new pattern elements through growth. The new elements formed by epimorphic regulation may be the same as those which were removed, in which case together with the regulating fragment they make up the complete pattern, or they may be the same as structures already present in the regulating fragment, leading to pattern duplication. French, Bryant & Bryant (1976) have recently developed a formal interpretation of epimorphic regulation based on the results of experiments on the legs of hemimetabolous insects, the imaginal discs of *Drosophila*, and amphibian limbs. This interpretation assumes that positional information in field which exhibit epimorphic regulation is specified in terms of polar co-ordinates where one component of positional information corresponds to position on a radius and the other to position on a circle (0-12 sequence). In amphibian limbs, the radial sequence is in the proximo-distal axis, and the outer circular sequence around the base of the limb. It is proposed that cells in such fields respond to surgical intervention according to the following principles.

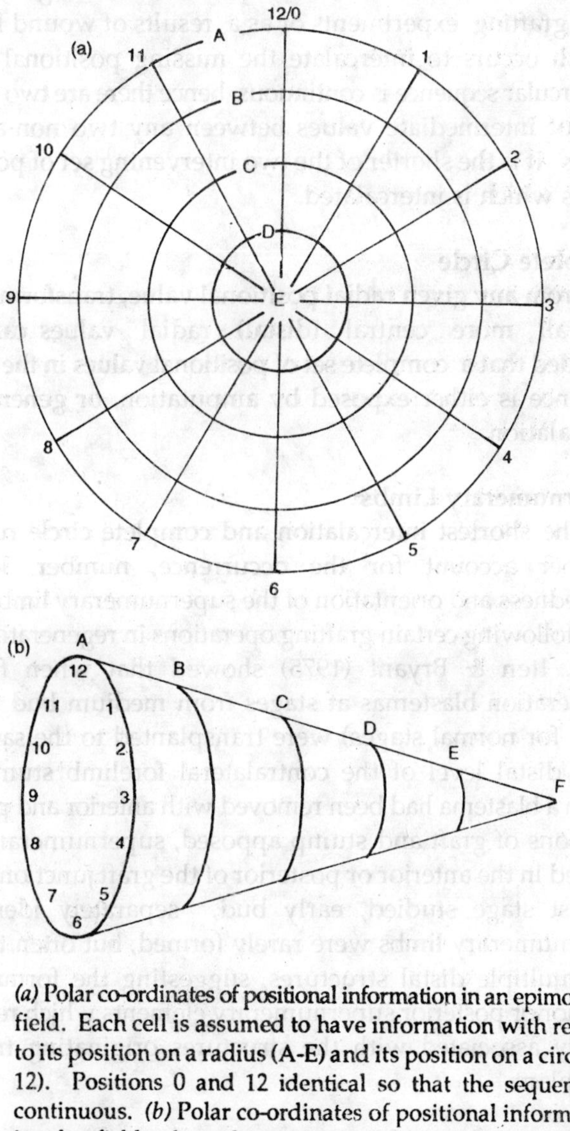

Fig. 8.1. *(a)* Polar co-ordinates of positional information in an epimorphic field. Each cell is assumed to have information with respect to its position on a radius (A-E) and its position on a circle (0-12). Positions 0 and 12 identical so that the sequence is continuous. *(b)* Polar co-ordinates of positional information in a leg field, where the most proximal extent of the field is represented by the largest circle, and the distal tip by the cone apex.

Shortest Intercalation

When cell with normally non-adjacent positional values in either the radial or circular sequences are brought together (as in grafting experiments or as a results of wound healing) growth occurs to intercalate the missing positional values. The circular sequence is continuous, hence there are two possible sets of intermediate values between any two non-adjacent values. It is the shorter of the two intervening set of positional values which is intercalated.

Complete Circle

From any given radial positional value, transformation to form all more central (distal) radial values can occur provided that a complete set of positional values in the circular sequence is either exposed by amputation, or generated by intercalation.

Supernumerary Limbs

The shortest intercalation and complete circle rules can together account for the occurrence, number, location, handedness and orientation of the supernumerary limbs which form following certain grafting operations in regenerating newt limbs. Iten & Bryant (1975) showed that when forelimb regeneration blastemas at stages from medium bud to early digits for normal stages) were transplanted to the same or a more distal level of the contralateral forelimb stump from which a blastema had been removed with anterior and posterior positions of graft and stump apposed, supernumerary limbs formed in the anterior or posterior of the graft junction. At the earliest stage studied, early bud, separately identifiable supernumerary limbs were rarely formed, but often the limb had multiple distal structures, suggesting the formation of anterior or posterior supernumerary elements which remained closely associated with the structures originating from the transplant.

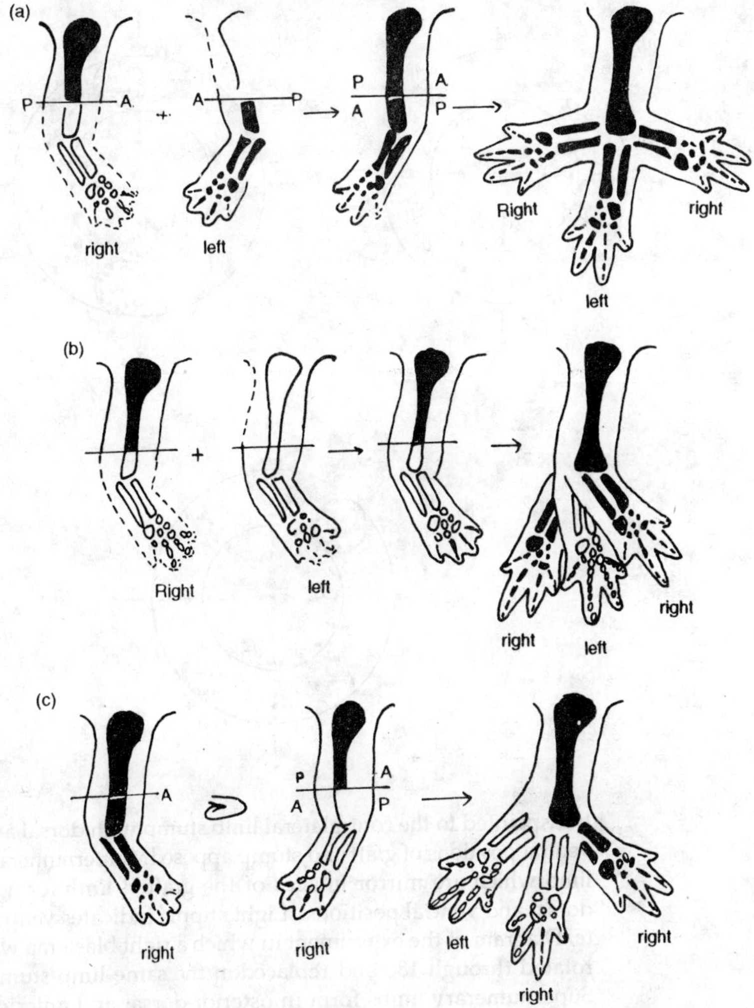

Fig. 8.2. (a) Diagram of the experimentation which a left blastema was transplanted to the contralateral limb stump with anterior and posterior positions of graft and stump apposed. Supernumerary limbs which are mirror images of the grafted limb form in anterior and posterior positions.

A anterior; P, posterior; black shading indicates dorsal *(b)* Diagram of the experiment in which a left blastema was

Fig. cont...

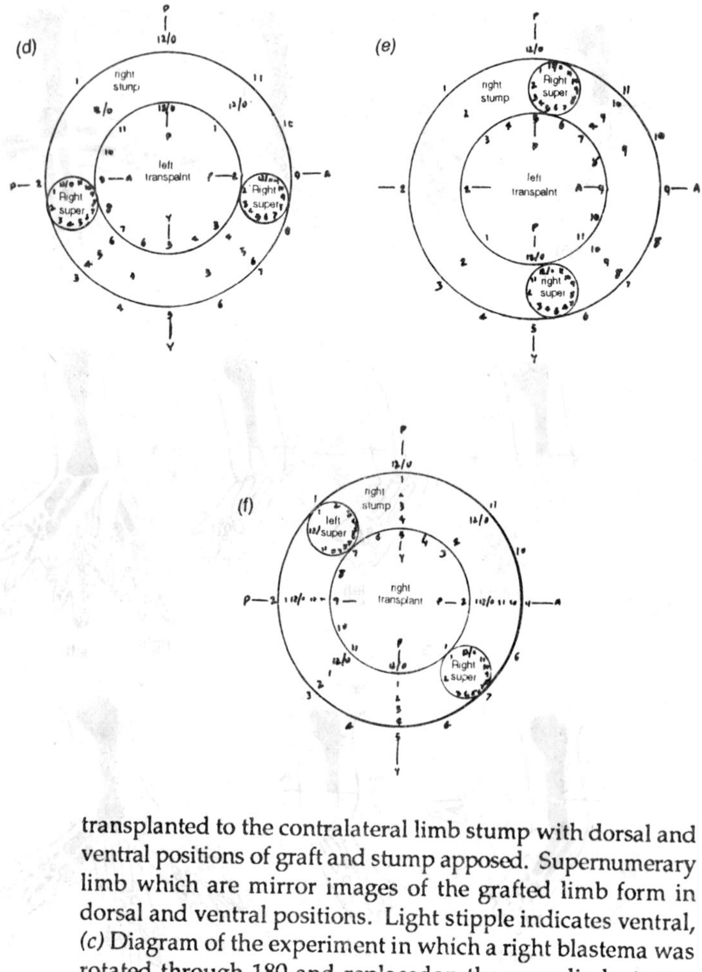

transplanted to the contralateral limb stump with dorsal and ventral positions of graft and stump apposed. Supernumerary limb which are mirror images of the grafted limb form in dorsal and ventral positions. Light stipple indicates ventral, *(c)* Diagram of the experiment in which a right blastema was rotated through 180 and replacedon the same limb stump. Supernumerary limbs form in osterior-dorsal and anterior-ventral positions. Light stipple indicates ventral. *(d)*, *(e)* and *(f)* Schematic cross-sections of graft/host junction distal view: outer circle, host circumference; inner circle, graft circumference. The diameters of graft and stump are shown to be different for clarity only. The circular sequence is indicated around the circumference by numbers 0-12. The numbers between the circles are values generated by shortest intercalation between different positional values of host and graft at the junction. The shorter route around the circle is different on the two sides of each point of maximum incongruity, so a complete circular sequence is generated at that position. Distal transformation occurs from these complete circles, to generate supernumerary limbs. The arrangement of positional values around the supernumerary limbs, and hence their orientation and handedness, is determined by the direction of intercalation in the adjacent regions of the graft junction, and is a consequence of the shortest intercalation rule. *(d)* Schematic cross-section of the graft/host junction following contralateral grafting with anterior and posterior positions apposed. Both supernumerary limbs are of stump handedness, they are oriented like the stump, and are mirror images of the graft. *(e)* Schematic cross-section of the graft/host junction following contralateral grafting with dorsal and ventral positions apposed. Both supernumerary limbs are of stump handedness, they are oriented like the stump and are mirror images of the graft. *(f)* Schematic cross-section of the graft/host junction following rotation through 180° and ipsilateral grafting. The posterior-dorsal supernumerary is of opposite handedness to the stump and graft, and is arranged in mirror-image symmetry with the graft. The anterior-ventral supernumerary is of the same handedness as both graft and stump, and is oriented like the stump. Super, supernumerary limb; A anterior; P, posterior; D, dorsal: V, ventral.

In a subsequent study, Bryant & Iten (1976) showed that dorsal or ventral supernumerary limbs would form following transplantation of a blastema at medium bud or early digits stages to the contralateral limb stump with dorsal and ventral positions of graft and stump apposed. Thus it appeared that supernumerary limbs could form wherever anterior and posterior or dorsal and ventral tissues were in contact. However, when both dorsal and ventral as well as anterior and posterior positions of graft and stump were apposed by ipsilateral grafting of a blastema rotated by 180°, a maximum of two supernumerary limbs were formed, not four as might have been expected on the basis of the results from contralateral grafts.

In both the contralateral and ipsilateral grafting operations the location, handedness and orientation of the supernumerary limbs are reproducible. When anterior and posterior positions

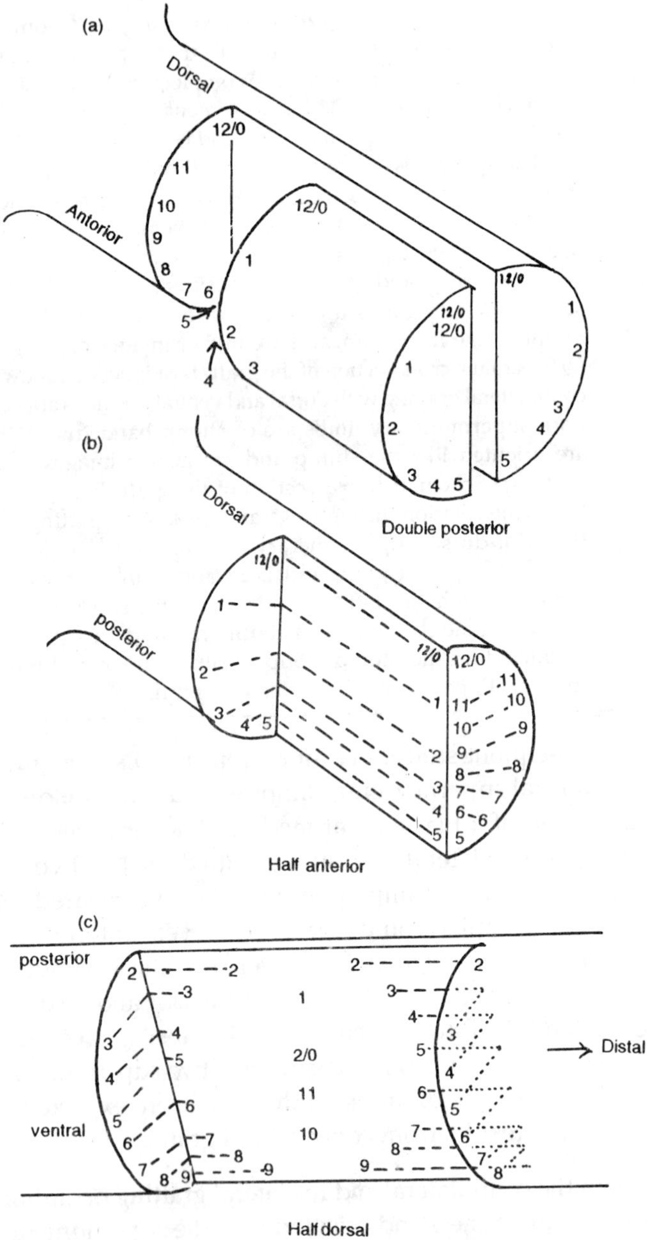
(a)
Dorsal
Antorior
12/0
11
10
9
8
7
6
5
4
12/0
1
2
3
12/0
12/0
1
2
3
4
5
12/0
1
2
3
4
5
Double posterior
(b)
Dorsal
posterior
12/0
1
2
3
4
5
12/0
1
2
3
4
5
12/0
11
10
9
8
7
6
5
11
10
9
8
7
6
Half anterior
(c)
posterior
ventral
Distal
2/0
11
10
1
Half dorsal

Fig. 8.3. (a) Diagram of a double posterior limb stump, made on the right upper arm. The graft is shown as being considerably smaller than the host-site in order that the proximal graft/host junction can be visualised. Positional values in the circular sequence are indicated on the host and graft. Values comprising a complete circle are present at the proximal graft/host junction.

---

were apposed in grafting, supernumerary limb arose in the anterior and posterior and were arranged in mirror-immage symmetry with the graft. Similarly, when dorsal ventral positions were apposed, the dorsal and ventral supernumerary limbs were mirror images of the graft. However, when both dorsal and ventral as well as anterior and posterior positions were apposed by rotation of the graft through 180°, supernumerary limbs formed either in a posterior-dorsal location or in an anterior-ventral location, with respect to the stump. The posterior-dorsal supernumerary limbs were mirror images of the grafted limb, whereas the anterior-ventral supernumerary limbs were of the same handedness as the grafted limb.

As has been pointed out by French *et al.* (1976) and Bryant & Iten (1976), all of these results can be understood as a consequence of the shortest intercalation and complete circle rules, if it is assumed that in the regenerating appendage positional values in the circular sequence are slightly non-uniformly distributed. Intercalation between graft and stump can occur where normally non-adjacent positional values in the circular sequence are juxtaposed. For each type of graft illustrated, there are two positions around the graft junction where the two possible routes of intercalation involves the same number of positional values. Intercalation on either side of these locations is in opposite directions, and a complete circle is generated where the two half-circumferences meet. Distal transformation can then occur from these complete circles, but not form any other position in the graft junction. The handedness and orientation of the supernumerary limbs so produced is controlled by the direction of intercalation in the regions of the graft junction adjacent to them.

Recent studies have shown that the ability to form supernumerary limbs is not limited to regenerating limbs. When mature, non-regenerating hand were transplanted to the contralateral limb stump, supernumerary limbs in the expected locations, and with the predicted handedness, sometimes formed. The incidence of supernumerary limb formation in mature limbs was greater if the graft was made to a more proximal level on the contralateral limb, rather than to the same level. The reason for this difference in incidence is not understood, but it may indicate that mature, fully differentiated tissues do not readily respond to discontinuities of positional values in a single sequence. When mature limb tissues were grafted to create double half-limbs (see below), a single supernumerary limb sometimes formed at the proximal junction between graft and limb stump. This is the position at which a complete circle of positional values is created by the grafting operation on. The fact that the formation of a supernumerary limb in this location was often not apparent until after the limb had been amputated through the distal region of the graft and dedifferentiation of the graft had been initiated, suggests that dedifferentiated cells may be better able to respond to discontinuities in positional values than are differentiated cells.

In addition to the regenerating and non-regenerating limbs of adult urodeles, the developing limbs of larval urodeles are also known to be capable of producing supernumerary limbs following certain types of orthotopic and heterotopic transplantations. The results of many of the pioneering experiments of Harrison and Swett as well as those of other workers, can be understood in terms of the polar co-ordinate model.

Double Half-Limbs

Further support for the complete circle rule comes from experiments on double half-limbs. To make double anterior or double posterior limb stumps, half of one upper arm was removed and replaced by the reciprocal half from the

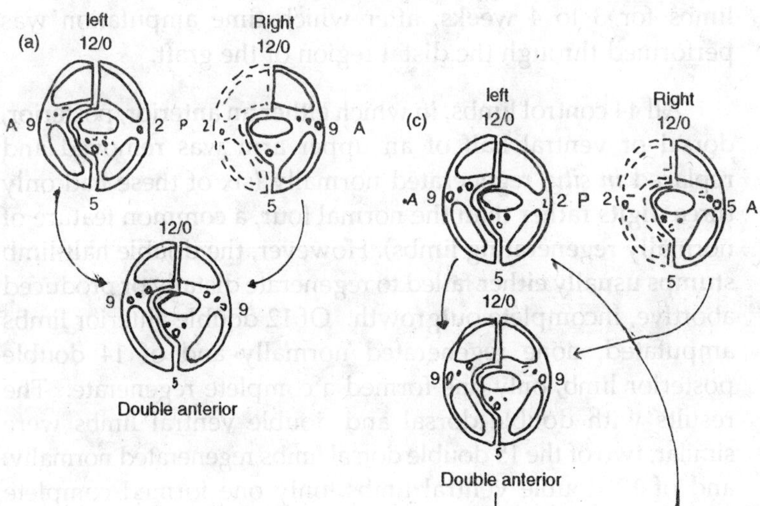

Fig. 8.4. Schematic diagram of operations to produce double half and reversed normal limbs. The upper arm is viewed in cross-section from a distal position. Key positional values are indicated on the limbs at dorsal, ventral, anterior.

contralateral upper arm. Similar operations were performed to create double dorsal and double ventral stumps. The grafted pieces were allowed to heal into place on the unamputated limbs for 3 to 4 weeks, after which time amputation was performed through the distal region of the graft.

All 44 control limbs, in which either an anterior, posterior, dorsal or ventral half of an upper arm was removed and replaced *in situ*, regenerated normally (six of these had only three digits rather than the normal four, a common feature of normally regenerating limbs). However, the double half-limb stumps usually either failed to regenerate distally or produced abortive, incomplete outgrowth. Of 12 double anterior limbs amputated, none regenerated normally and of 14 double posterior limb, only one formed a complete regenerate. The results with double dorsal and double ventral limbs were similar; two of the 15 double dorsal limbs regenerated normally, and of 12 double ventral limbs, only one formed complete distal regenerates. Histological examination of double half-limb stumps at the time of amputation showed that they were innervated, and comparison with control limb stumps in which half of the upper arm had been removed and replaced, revealed no major differences in appearance which might account for the difference in regenerative ability. In addition, double half-limb of all types initiated regeneration, and several proceeded as far as the stage of late bud. Histological examination of double half-limbs during these early stages of regeneration shows that dedifferentiation, blastema accumulation and formation of a thickened apical cap all occur as normal.

Regenerative failure in these limbs is not due to an inadequate nerve supply, a shown by the control limbs, which regenerate normally, and the histological evidence cited above. Further evidence come from two sets of experiments. First, eight double half-limb stumps in which the normal interval between grafting and amputation was doubled to 2 months to allow extra time for reinnervation, also failed to regenerate. Secondly, in a double grafting operation nine double half-limbs were made, allowed to heal and become vascularised for 3-4 week, and then converted to reversed normal limbs by replacing the originally ungrafted half-limb with a reciprocal half from the contralateral limb. The reversed normal limbs were allowed an interval of 3-4 weeks prior to amputation. Such limbs formed normal regenerates, with opposite handedness to that of the original limb, as expected from the reversed normal stump.

The regenerative failure of the double half-limbs provides direct support for the complete circle rule. Regeneration of a few of the double half-limbs may have been the result of the formation of complete circles from one or both half-limbs by intercalation in the circular sequence.

Further support for the complete circle rule can be found in the experiments of previous investigators. X-irradiated amphibian limbs will not normally regenerate when amputated. However, Lheureux and Carlson (1974) have shown that an X-irradiated limb stump will regenerate if it is provided with a normally oriented cuff of non-irradiated skin. Regeneration does not occur if a cuff of skin is taken from one surface of the limb and is oriented so that only one 'quality' (dorsal, ventral, anterior or posterior) is present at the amputation site. These results suggest that regenerative ability can be restored in X-irradiated limbs by a skin cuff containing all positional values in the circular sequence, but cannot be restored if only part of the circular sequence is present.

Half-Limbs

Previous studies have shown that longitudinal half-limbs can regenerate normally or relatively normally. In parallel with the double half-limb experiments described in the previous section, we have also looked at the regenerative ability of half-limbs stumps under two different conditions: first, removal of half of the upper arm, followed by immediate amputation through the distal region of the half-limb; secondly, as above, but with amputation 1 to 2 months after the removal of half the limb.

Of the 12 half posterior limbs amputated immediately, 10 formed normal regenerates (two had only three digits), and two failed to regenerate. Of the 12 half anterior limbs amputated immediately, 10 regenerated normally (one had only three digits) and two failed to regenerate. The regenerates originated over the junction between the distal and lateral would surfaces, but during the later stages of regeneration most became more normally oriented on the distal half-limb stump. A tentative explanation for these results is that the missing positional values can be restored along the lateral wound surface by migration of cells during wound healing from the proximal half-limb stump. A complete circle of positional values can therefore be created at the edge between the distal and lateral wound surfaces, and distal transformation can occur from this complete circle. A similar explanation can account for the results of experiments in which half-blastemas were removed at different stages of regeneration. At all stages prior to early digits, such half-blastemas developed into normal limbs. An explanation for this result is that the remaining half-blastemas, together with cells migrating during healing from the proximal stump, will constitute a complete circle of positional values, leading to the formation of a normal regenerate.

When amputation of half-limbs is performed 1 to 2 months after removal of half of the limb, the results differ in some respects from those following immediate amputation. Of 20 half posterior limbs amputated after 1 or 2 months, 17 regenerated normally (two had only two digits and one had three digits). A further three half posterior limbs formed a regenerate from the distal wound surface, and supernumerary elements from the lateral wound surface, leading to a total of five or six digits per half-limb. Of 19 half interior limbs amputated after 1 to 2 months, 17 regenerated normally and two failed to regenerate. All of the 14 half ventral limbs amputated after 1 month regenerated normally. Of 14 half dorsal limbs amputated after 1 month, 10 regenerated normally (two had less than four digits) and three formed a regenerate from the distal wound surface, and additional structure from the lateral wound surface. The remaining limb formed structures from the lateral wound surface but not from the distal amputation surface. The formation of a normal regenerate from the distal surface. The formation of a normal regenerate from the distal surface of the half-limb can be understood if it is assumed that the missing positional values are replaced along the distal region of the lateral wound surface by migration during wound healing from the distal wound surface. The occurrence of multiple regenerates on some half posterior and half dorsal limbs but not on half anterior and half ventral limbs, suggests the possibility that shortest intercalation may occur in the central regions of the long lateral wound prior to wound healing. If positional values around the circle are slightly non-uniformly spaced, as we have suggested from the supernumerary limb results, then shortest intercalation on the lateral wound surface of a half ventral or a half anterior limb stump will not introduce any positional values which are not already provided by migration from the distal and proximal wound surfaces. Hence, complete circles of positional values will not form on the lateral wound surfaces of such half-limbs. However, shortest intercalation on the lateral surface of a half dorsal or half posterior limb stump could introduce new positional values which are not provided by migration and could lead to the formation of two complete circles on the lateral surface. Distal transformation could then occur from one or the other of these complete circles. The fact that the structures forming on this surface can sometimes be recognised as being of the same handedness as the normal distal regenerate, and sometimes of opposite handedness, supports this suggestion. The opportunity for intercalation in these half-limbs may depend on the rate of wound healing from the proximal and distal stumps. Half-limbs with delayed amputation will have a greater opportunity for intercalation compared to half-limbs amputated immediately, because the length of the lateral wound is greater in the former than in the later. In the latter limbs, the size of the lateral wound is reduced due to amputation, wound trimming and subsequent dedifferentiation of the stump.

The relationship of wound healing to the results obtained with half-limb stumps is being explored further, using scanning electron microscopy.

9

Nerves and Limb Regeneration

Neurotrophic function or effects have been defined a those interactions between nerves and other cells which initiate or control molecular modifications in the other cell' ... These interactions are, for the most part, long-term ones. This rather broad definition embraces trophic effects in vertebrate regeneration as well as trophic effects involved in the maintenance and function of skeletal muscle organs and transneuronal circuits in the central nervous system (CNS). The considerable body of information on the influence of nerves in tail regeneration is beyond the scope of this review.

The Nerve Requirement in Amphibian Limb Regeneration

The first clear suggestion that nerves were important in the regeneration of vertebrate appendages came from the studies of Todd in 1823. Almost a century intervened before Wolff (1902) reaffirmed Todd's view of the importance of nerves in salamander limb regeneration. Studies in the 1920s served to emphasise the obligatory nature of the nerve requirement. Once the importance of nerves was generally accepted, there followed a literature suggesting that the neural influence was a qualitative one. The question most frequently asked was: Which of the various types of nerves provides the trophic influence? Each functional division of the nervous system, i.e. motor sensory and sympathetic had its champions.

We now know, as a result of a series of classical studies executed by Marcus Singer that limb regeneration in amphibians is dependent upon the *quantity* of nerve tissue at the amputation surface of the limb. Motor, sensory and, presumably, sympathetic fibres are all, if present in sufficient numbers, equally capable of supplying the essential neurotrophic factor. Initially this relationship between nerve and limb tissue was expressed as the threshold number of Bodian-stainable nerve fibres per 100 µm2 of amputation surface required for regeneration at the level of the forearm of the forelimb in *Triturus*.

This quantitative relationship proposed for regeneration in adult organisms is not unique to adult forms, but has been tested and found to hold also for limb regeneration during the three post-embryonic stages of development in *Triturus* as well as for post-embryonic stages of anuran development. The latter case is especially instructive, since larval anurans gradually lose their ability to regenerate limbs as they approach the metamorphic climax. Van Stone (1964) found that the number of nerve fibres per unit area of stump tissue decreased in accordance with limited regenerative ability in the developing tadople of *Rana sylvatica*. Moreover, the normally non-regenerating limbs of adult vertebrates such as *Rana* lizards and mice were found to have nerve fibre numbers well below the threshold number required for regeneration of adult *Triturus* limbs. These correlative bits of evidence are strengthened by the fact that in *Rana*, three lizard species and the neonatal opossum, augmentation of the normal neural component evokes a regenerative response in these normally non-regenerating limbs. An apparent exception to the threshold fibre requirement appeared in the form of the South African clawed toad, *Xenopus laevis*. Adult *Xenopus and Hymenochirus*, members of the family Pipidae, unlike other anurans, exhibit definite (albeit heteromorphic) limb regeneration. Determination of the number of nerve fibres per unit area of amputation surface in the limb of *Xenopus* revealed that the innervation level was not only below the threshold level of the *Triturus* limb, but was lower than the values for the normally non-regenerating limbs of the lizard and even the mouse. However, *Xenopus*, like other forms capable of regenerating limbs, does require an adequate nerve supply. Rzehak & Singer (1966) noted that although *Xenopus* had relatively few nerve fibres per unit area of tissue, the diameter of many of the nerve fibres exceeded that of the other animals that they had studied. Thus, they reasoned that larger axons might well be more effective carriers of the trophic influence. They found that if fibre diameter was taken into consideration and the amount of neural substance at the amputation surface was expressed as the area of axoplasm per too μm^2 of amputation surface, then *Xenopus* no longer presented a paradox. For, when neural contribution is expressed in this manner, the limbs of *Triturus* and *Xenopus* are essentially equivalent.

Regeneration of Aneurogenic Limbs in Larval Salamanders

Having just emphasised the generality of the neurotrophic hypothesis of regeneration, we must now grapple with the intriguing fact that aneurogenic and/or *sparsely innervated* limbs of larval *Ambystoma* regenerate quite well. This was first demonstrated by Chester Yntema (1959*a*, *b*) who removed the neural tube and associated neural crest from tailbud stage embryos of *Ambystoma maculatum*, parabiosed them to normal tailbud embryos, and reared the parabionts to advanced larval stages. The operated member of the parabionic pair, nurtured by its normal partner, developed and produced limbs that were, by silver staining techniques, determined to be either aneurogenic or in some cases sparsely innervated. Upon amputation these limbs regenerated. This result was particularly surprising since previous studies had demonstrated that the regeneration of normally innervated larval salamander limbs was, in some respects, even more dramatically dependent upon proper innervation than was the adult limb. Whereas amputation and denervation of the adult limb prevents regeneration and leads to wound healing and callus formation, amputation of the denervated larval limb was shown to result in continued dedifferentiation and ultimately to partial or complete resorption of the remaining limb stump. A note of caution must be interjected at this point. Based on the studies of Egar & Singer (1971), it is quite possible that *all* of the limbs produced by the Yntema technique are to some degree sparsely innervated and that such limbs designated as sparsely innervated in reality possess more nerve fibres than the classical Bodian silver stain would reveal. This technicality does not seriously alter the interpretation that have subsequently been put upon studies of the 'aneurogenic' system, for the neurogenic hypothesis of regeneration rests largely, if not completely, on the same criterion, namely that of Bodian-stainable nerve fibres. Singer (1952), in an attempt to account for the dispensability of nerves in regenerating aneurogenic limbs, offered the following provocative and, as it turned out, discerning hypothesis:

> It would seem that the nerve dependence of regeneration of the limb arises as a result of association with the nerve during embryonic development. Stated otherwise, the limb in the course of its development becomes 'addicted', so to speak, to the nerve contribution. Or, perhaps, in the absence of the nerve, other growth factors, with which the embryo abounds but which may diminish later, are preserved in greater abundance in developing nerveless limbs and thereby compensate for the absence of nerve stimulation.

If, in fact, limbs do become 'addicted' to nerves, adult salamander limbs appear not to respond to prolonged withdrawal (denervation) therapy. However, in a series of studies on aneurogenic limbs of *Ambystoma,* Charles Thornton and his colleagues laid the groundwork for testing the 'addiction' hypothesis. This system is particularly tailored for testing the hypothesis, since nerves are absent during the development of the aneurogenic limbs; however, nerves could be introduced by orthotopic transplantation of the aneurogenic limb to normal larvae where brachial nerves then invade the graff. Moreover, the new innervation can they easily be withdrawn. Thornton

& Thornton (1970) published an elegant series of experiments, the results of which clearly demonstrated that within 13 days after orthotopic transplantation, innervation of the aneurogenic limb grafts rendered them nerve-dependent with respect to regeneration. They further demonstrated that when the innervated (nerve-dependent) aneurogenic limb grafts were deprived of their newly acquired nerves and maintained in a denervated state for 40 days prior to amputation, as many as 49% had regained their original ability to regenerate in the absence of nerves. Thus, these results are consistent with the 'addiction' hypothesis of Singer and suggest that over short time intervals during larval life, the nerve dependency ('addiction') of the limb tissues is reversible.

The skin, ultimately the wound epithelium derived from it, determines whether or not aneurogenic limbs will regenerate. These studies have shown that a limb stump constructed from skin of an aneurogenic limb and denervated (previously innervated) cartilage and muscle will regenerate. The reverse combination does not result in a regenerate. These observations are particularly intriguing when one considers the important role of the wound epithelium and apical epidermal ridge in normal limb regeneration and limb development.

Regeneration under sparsely innervated conditions is not unique to the aneurogenic limbs of larval salamanders. Studies on lizard tail regeneration have established the primacy of the ependymal epithelium (a non-neuronal cell type) as the source of the trophic influence in this system. Considerable support for the role of the ependyma in salamander tail regeneration has also been provided. Additionally, implants of ependyma placed in the normally non-regenerating limbs of lizards elicited regenerative responses indistinguishable from those previously induced by augmentation of the nerve supply.

The Nature of the Trophic Agent(s)

Having presented evidence for a neuronal influence on regeneration, we must now ask : What is the nature of the influence ? At present, the most commonly held hypothesis states that : (1) the neurotrophic agent(s) is molecular in nature and (2) the trophic substance(s) is produced in the perikaryon transported down the axon via axoplasmic transport and discharged from nerve endings to influence the regenerating tissues. Since no one has yet isolated a neurotrophic molecule, we at present accept the existence of such a molecule on the basis of several pieces of indirect evidence. Guth (1969), focusing primarily on nerve-muscle trophic influences, has emphasised the indirect nature of our evidence by referring to the putative trophic molecules as *mysterines*.

With respect to amphibian limb regeneration, the experiment of Kamrin & Singer (1959) demonstrated that isolated sensory ganglia implanted beneath a denervated early blastema would maintain the blastema and, ultimately, support regeneration. This, in and of itself, suggested the presence of a diffusible trophic molecule and clearly demonstrated that normal synaptic relationships were not essential. More striking evidence for a diffusible trophic molecule has been obtained in studies of nerve muscle and retina-iris interactions. Several studies have established that the maintenance of cholinesterase activity of the motor end plate in mature skeletal muscle is dependent on innervation. When muscle fibres are denervated, there is a reduction and eventual loss of cholinesterase activity at the motor end plate. It has been shown that this trophic relationship can be duplicated and studied in organ cultures of muscle and nerve. The effect maintenance of end plate cholinesterase in denervated organ-cultured skeletal muscle from *Triturus* by co-culturing the muscles with : (1) isolated sensory ganglia, *either separated from or in contact with* the muscle; and (2) homogenates of brain tissue added to the culture medium. It should be noted that *sensory ganglia* were able to support a function normally provided by motor neurons. Additional evidence for a diffusible trophic molecule comes from the elegant experiments of Yamada and his colleagues. In vivo studies have demonstrated that regeneration of a lens from the dorsal iris of lentectomised eyes of *Triturus* is dependent upon a trophic contribution from the neuroretina. This system has also been studied *in vitro*. Although organ cultures of dorsal iris will initiate some of the steps leading towards lens regeneration, frank lens formation does not occur. However, when portions of the dorsal iris are co-cultured with neuroretina from frog larvae pituitary glands or sensory ganglia complete lens regeneration is obtained in a large percentage of the cases. In this reviewer's opinion, these two in-vitro systems provide strong evidence that trophic influences are mediated by one or more diffusible 'trophic' molecules.

Given a putative trophic molecule, what evidence do we have concerning the transport of such molecules and their transfer to the tissues at the wound site ? We have excellent evidence for the transport of molecules via the axon, i.e., axoplasmic transport. That some of the molecules transported via nerves are in fact trophic has been suggested by studies on nerve-muscle preparations. However, evidence for *specific* transfer of materials from axon terminals to recipient cells is limited. Evidence has been presented for the movement of labelled material from motor nerves into muscle cells and from the retina to the striate cortex, via the lateral geniculate.

Considerable effort has been expended by a number of laboratories in an attempt to isolate and identify a trophic molecule. Historically, neurotransmitters were judged to be the most likely candidates. Acetylcholine, in particular, has long enjoyed favour. Several lines of reasoning, however, argue against neurotransmitters being the trophic molecule. Sensory nerves alone are capable of supplying the trophic influence in limb regeneration, and apparently suffice for maintaining acetylcholinesterase at motor end plates, as well as lens regeneration from the dorsal iris in *Triturus*. Obviously the direction of nerve impulse propagation and neurotransmitter release in sensory nerves are in the opposite direction to the trophic influence. This argument does not, however, rule out the possibility that the trophic agent may be co-released with acetylcholine in nerve-muscle trophic interactions. Singer (1960) was able to show that the acetylcholine content of *Triturus* limbs innervated solely by motor nerves was much higher than that of limbs possessing only sensory innervation. Yet the former did not regenerate, while the latter did. Although newt limb stumps infused with a variety of chemicals known to block the acetylcholine mechanism in various ways often fail to regenerate, the action of these chemicals appears to be due to general toxic effects rather than their specific blockade of the acetylcholine mechanism. Neither the application of acetylcholine, nor chemicals that are known to accentuate, prolong, mimic or protect acetylcholine, appear to substitute for the trophic agent. Finally, Drachman & Singer (1971) have shown that complete blockade of acetylcholine transmission with botulinum toxin does not inhibit limb regeneration in *Triturus*. The results of this study are most compelling. Botulinum toxin is the most specific and potent cholinergic blocking agent known. Thus, we may conclude that for limb regeneration acetylcholine is neither necessary nor sufficient. Likewise, the possibility that noradrenaline (norepinephrine) is the trophic agent is ruled out by the fact that sympathectomised newt limbs show no impairment of regeneration, nor do limb stumps innervated solely by sympathetic fibres regenerate. The results derived from these studies are compatible with those mentioned earlier, which demonstrated that a non-nerve cell type, ependyma as well as non-neuronal tissues of aneurogenic salamander limbs can apparently provide the trophic agent necessary for regeneration. Taken together these studies argue against the trophic molecule(s) being unique to the neuron and have served to direct our search away from molecules that would be unique to nerve tissue.

The ability to isolate and characterise a biologically active molecule depends heavily on the availability of a rapid, sensitive and highly specific bioassay system. To date, most attempts to isolate the trophic molecule have utilised a bioassay developed by Singer, Weinberg & Sidman (1955). In this bioassay, moderate-early to early blastema stage regenerates of *Triturus* are denervated. Regenerates denervated at this stage cease regeneration and the young blastema shrivels and disappears. Forty-eight hours following denervation, tissue homogenates or chemicals to be rested are then microinfused via a fine glass needle inserted into the limb and positioned directly beneath the young blastema. In most studies, the end point of the bioassay has been maintenance of the blastema and/or continued limb regeneration. If the results of this bioassay are to be meaningful, it must be rigorously demonstrated : (1) that denervation was complete; and (2) that the denervated state was maintained during the test period. Using this procedure, homogenates of regenerating blastemas nerve, brain and liver extracts of spinal cord and brain and various chemical substances have been tested for their ability to support, grossly and or histologically, visible regeneration. To date, none of these attempts have succeeded in supporting limb regeneration.

Several studies, employing the same procedure but relying on a potentially more sensitive end point for the assay, have yielded positive and therefore more interesting results. The end point chosen in these studies was the maintenance of macromolecular synthesis in the denervated early regenerate. When early regenerates are denervated, protein synthesis in the young blastema is reduced by 40% when compared to the contralateral innervated blastema. This suggests that approximately 40% of protein synthesis in the early regenerate is under neurotrophic control, a figure close to that found for nerve-dependent synthesis of muscle cholinesterase. Lebowitz & Singer (1970) also showed that the reduction in protein synthesis was proportional to the extent of the denervation. Singer & Caston (1972), in a similar study, found that denervation was followed by an initial burst of macromolecular synthesis (RNA, DNA and protein) followed by a decrease of 50-60% compared to the innervated control. They suggested that the nerve influences either synthesis, processing or turnover of macromolecules. RNA synthesis has been followed in innervated and denervated larval axolotl limbs following denervation. Here RNA synthesis increased in both, but a greater increase was seen in innervated limbs. DNA synthesis was found to follow the same pattern.

Infusion of denervated early regenerates with homogenates or supernatants of nerves taken from regenerating *Triturus* limbs restored some 60% of the deficit in protein synthesis induced by denervation. Synaptosomal concentrates from frog brain and newt brain extracts infused in a similar manner resulted in values comparable to or exceeding control-innervated or sham-infused values. In this same study infusions of liver and spleen homogenates or the mitochondrial fraction of brain homogenates led to [^{14}C] leucine incorporation levels that fell within the range of denervated controls. The stimulatory effect of brain extract was destroyed by trypsin digestion and by heating (60 °C for 10 min). Singer *et al.* (1976) have suggested that the active agent is a protein of between 200 000 and 300 000 molecular weight. Since there is no significant change in the rate of protein degradation and no rate-limiting change in the amino acid pool, these authors have suggested that the neurotrophic agent predominantly influences synthesis of proteins in the blastema. Dearlove & Stocum (1974) have analysed by acrylamide gel electrophoresis the proteins of regenerates before and after denervation. They found that three protein bands disappeared following denervation, and suggested that these three bands represented nerve-dependent proteins.

The infusion experiments of Lebowitz & Singer (1970) and Singer *et al.* (1976) represent the best evidence to date that the neurotrophic agent in amphibian limb regeneration is chemical in nature, and offer encouragement that infused, denervated early regenerates can be used as an effective bioassay in future attempts to isolate and purify the trophic molecule(s).

Nature of the Tissue Response to the Trophic Agent

Remarkably little effort has been directed towards

elucidating the response(s) of the wound tissues and blastema to the trophic agent. Although the studies discussed under the previous heading have demonstrated that macromolecular synthesis in the early regenerate is influenced by the trophic agent, such a response on the part of the blastema is a very general one. This response is in fact not unique to the early regenerate, where the trophic agent exerts an all-or-none effect with respect to regeneration, but is exhibited even by advanced regenerates. Although denervation at later stages of regeneration does not seriously impair morphogenesis of the regenerate, the volume of the regenerate, like that of the mature limb, is at all stages nerve-dependent. In *Triturus* limbs between days 10 and 13 following amputation, the trophic agent is crucial if regeneration is to ensue. At this time, the early regenerate consists of a well formed wound epithelium and a small accumulation of rapidly proliferating blastema cells. Denervation at this stage effectively blocks continued growth of the blastema, which shrinks in size and is converted into scar tissue. Mitoses in the young blastema are profoundly reduced in number or completely blocked when the trophic agent is withdrawn in these early regenerates. Thus one of the prime functions of the trophic agent may well be that of maintaining cell proliferation in the blastema. This is supported by the fact that DNA synthesis in early blastemas is depressed by 50 to 60% within 48h following denervation. The relationship between the neurotrophic agent and blastema cell proliferation is further emphasised by the studies of Kelly & Tassava (1973) and Tassava *et al.* (1974). They followed the mitotic index in innervated and denervated limbs of larval axolotls for 7 days following amputation. Here, denervation and amputation were performed simultaneously. They found that the mitotic index increased over the 7-day period, but only in the innervated limbs. Tassava *et al.* (1974) also fund that DNA synthesis occurred, albeit at a reduced level, in denervated limbs, though few if any cells underwent mitosis in these limbs. They proposed that denervation results in blockage of cells in the G_2 phase of the cell cycle. Since denervation of larval limbs leads to resorption of the wound tissues, presumably the G_2-blocked cells degenerate and are removed. It is possible that this same sequence of events may follow denervation in regenerating adult salamander limbs as well, for, as we have seen, denervation leads to a decline in DNA synthesis and mitotic activity. Moreover, Bryant, Fyfe & Singer (1971) have described, at the ultrastructural level, changes in the blastema cell population occurring within 3 days of denervation. They observed a decrease in the number of common blastema cells and a concomitant increase in two cell types (pale cells and vacuolated cells) that were not found in innervated blastemas. In some cases they observed cells that were ultrastructurally intermediate between the normal blastema cells and these unique cell types. One is tempted to speculate that the pale and vacuolated cells observed in their study may represent degenerating G_2-blocked blastema cells.

A neurotrophic influence on cellular proliferation has also been implicated in the maintenance of taste buds in the frog lens regeneration from the dorsal iris in newts and modulation of epidermal mitotic activity. This aspect of the neurotrophic influence deserves careful analysis in future studies.

10

Limb Regeneration and Tissue Interactions

The central question of pattern formation during regeneration of the amphibian limb is how distal transformation of blastema cells occurs to reconstitute the missing structures. The limb regeneration blastema has a high degree of self-organisational capacity along its proximal-distal (PD) axis; therefore, the PD pattern of the regenerate would seem to be organised by the action of intra-blastemal factors rather than by the adjacent limb stump. The mechanism of this organisation is not understood, but it has been known for sometime that the blastema epidermis and limb nerves are indispensable requirements. The epidermal requirement for regeneration is continuous, the epidermis being essential for both blastema formation and the later outgrowth and morphogenesis of the blastema. The neural requirement is continuous or discontinuous, depending on what aspects of regeneration are being examined. Nerves are essentia for the formation of a blastema and for it continued growth in volume, but an established blastema becomes progressively independent of the nerve for its morphogenesis. Denervation reduces DNA synthesis and mitotic activity of the blastema. It has been proposed the epidermis functions to maintain as apical zone of blastema cells in an undifferentiated condition, and that these cells require a neural influence in order to traverse the G_2 portion of the cell cycle. The zone of undifferentiated blastema cells is of constant width and corresponds to the apically undifferentiated blastemal mesenchyme seen in histological preparations.

Recently, Iten & Bryant (1975) and Stocum (1975b) have demonstrated the inercalary regeneration of missing intermediate parts when a blastema is grafted to a more proximal segment of the limb stump. The intercalated parts appear to be formed from stump-derived blastema cells and not from the grafted blastema. Such intercalary blastema cells supposedly would be outside the normal zone of influence of the graft epidermis, and the possibility thus arises that this type of regeneration might require nerves, but not epidermis. This would be important, because if true it would be evidence against the idea that the epidermis functions to induce mesenchymal outgrowth or acts in other ways on pattern formation, such as in the progress zone model of distal transformation.

Here I will describe experiments that were undertaken to find out if intercalary regeneration is subject to the same neural and epidermal requirements as distal regeneration. The experiments were not altogether successful in answering this question with regard to the epidermis, but did reveal an unexpected feature of intercalary regeneration that is of potential importance, and provide a stimulus for discussion of the possible roles the epidermis might play in regenerate outgrowth and pattern formation.

Experiments

The degree of dependence of intercalary regeneration on nerves and epidermis was assessed by three different types of experiments in which either the epidermis, the nerve, or both were prevented from interacting with the regenerate. All experiments were made on larvae of *Ambystoma maculatum* (25 -40mm snout-tail length) and *Ambystoma tigrinum* (45-60mm snout-tail length). Larvae were first anaesthetised in 1 : 1000 MS:222 W/V, then transferred to 1: 10000 MS : 222 made up in full-strength Holtfreter solution for operation and healing under low-level anaesthesia. The staging system for the wrist regenerates used in these experiments has not been published. The following are brief descriptions of the stages used.

(1) *Medium Bud (MB).* Cytologically undifferentiated regenerate, conical in outline from a dorsal and ventral view but slightly flattened in the dorsal-ventral plane. Mean length/width (L/W) ratio = 0.46.

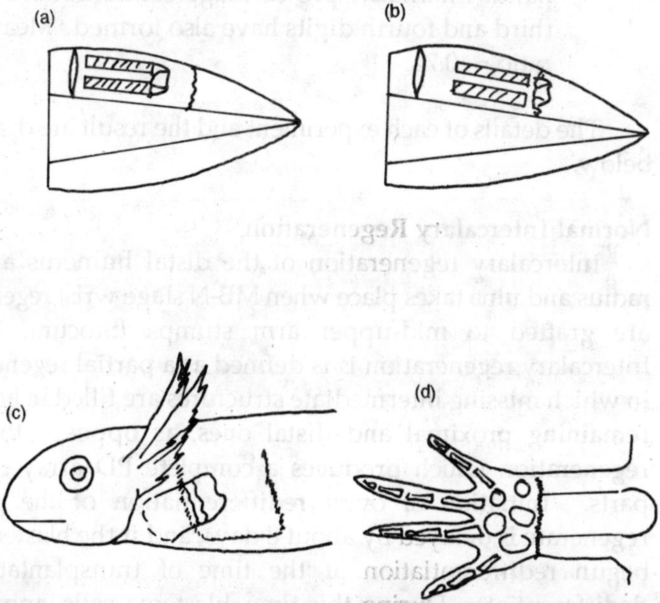

Fig. 10.1. **Diagram of the basic operations.** ***(a)*** **Wrist regenerate melded to upper arm stump in dorsal fin tunnel; no epidermis, no nerves.** ***(b)*** **Same operation except that the distal half of the regenerate projected from the tunnel and was re-covered by epidermis.** ***(c)*** **Epidermis-free wrist regenerate grafted to innervated upper arm stump, then inserted under a flap of flank skin.** ***(d)*** **Fully differentiated hand grafted to innervated upper arm stump.**

(2) *Late Bud (LB).* Cytologically undifferentiated regenerate with essentially the same external configuration as the MB, but with the mean L/W ratio increased to 0.56.

(3) *Early Redifferentiation (ER).* The external configuration is much the same as previous stages, except that the tip of the regenerate may be squared off. Pro-cartilage

condensations representing the first and second digits are now visible within. Mean L/W ratio = 0.69.

(4) *Notch (N).* An indentation has appeared at the apex of the regenerate, marking out the first two digits of the hand. Internally, pro-cartilage condensations of the third and fourth digits have also formed. Mean L/W ratio = 0.74.

The details of each experiment and the result are discussed below.

Normal Intercalary Regeneration

Intercalary regeneration of the distal humerus and the radius and ulna takes place when MB-N stage wrist regenerates are grafted to mid-upper arm stumps (Stocum, 1975b). Intercalary regeneration is is defined as a partial regeneration in which missing intermediate structures are filled in between remaining proximal and distal ones, as opposed to distal regeneration which produces a complete PD array of limb parts. Initiation of overt redifferentiation of the grafted regenerate is delayed by about 8 days, and if the blastema has begun redifferentiation at the time of transplantation, it dedifferentiates. During this time, blastema cells, apparently derived by dedifferentiation of the upper arm stump, accumulate proximal to the grafted blastema, approximately doubling the length of the regenerate. The portion of the total regenerate derived from the stump and that derived from the graft can be distinguished by the presence of an epidermal 'suture groove' at the junction where graft and stump healed together, and by differences in cell density. At 8 days post-operation, redifferentiation of the humeral and wrist and hand skeleton begins, followed by that of the radius and ulna. Preliminary labelling experiments, as well as histological studies, indicate that the wrist and hand redifferentiate from the grafted blastema, while the humerus and radius and ulna come from the stump-derived blastema cells.

Intercalary Regeneration in the Absence of Both Nerve and Epidermis

Forelimbs with MB-N stage wrist regenerates were served at their junction with the body wall and stripped of epidermis with watch-junction with the body wall and stripped of epidermis with watch-makers forceps after a 20-min treatment in calcium-free Holtfreter solution containing 0.2% EDTA. The regenerate and the proximal two-thirds or half of the upper arm were removed from the limb and melded together in normal PD orientation inside a tunnel bored in the connective tissue of the dorsal fin. The dorsal fin is innervated mainly by sensory nerves to the skin with few or no nerves in the connective tissue. This procedure effectively deprived the implants of both epidermis and innervation. Some specimens were fixed in Bouin's at 2-day intervals up to 10 days after the operation, serial sectioned at 10 μm and stained by the Holmes silver nitrate method; the remaining cases were stained *in toto* for cartilage by the Van Wijhe method, after a suitable period of development. Controls consisted of: (1) fin implants of wrist regenerates served and replaced at their original level; (2) fin implants of wrist blastemas alone; and (3) regenerates allowed to continue development *in situ* (isochronous controls).

The results of the cases stained for cartilage and prepared as whole mounts. The wrist blastemas differentiated carpal and digital skeletal elements in nearly all cases. The most distal elements regenerated were metacarpals, indicating the necessity of further epidermal contact for the outgrowth of phalanges. The important result, however, was that intercalary regeneration of the radius and ulna never occurred in DP implants, but in 16 of 27 cases (59.1%) some regeneration of the distal one-half or lone-third of the humerus did occur. This regeneration varied from a simple bridge of cartilage between the humerus and the basal carpal to a relatively well-formed humerus.

Histological examination of DD (17 cases) and DP (27

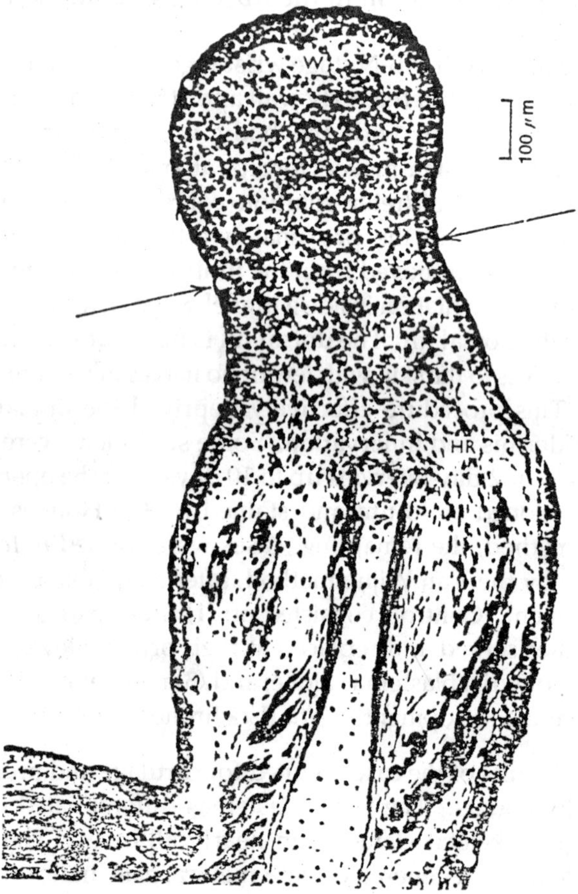

Fig. 10.2. Section through a normal intercalary regenerate 8 days after the operation. The arrows mark the dividing line between the granted wrist blastema cells (W) and the cells derived from the upper arm stump (1). Note the signs of cartilage dedifferentiation at the tip of humerus (H) and the beginning redifferentiation of the humeral regenerate (HR).

cases) implants revealed that the blastema redifferentiated muscles as well as cartilage, but in much sparser amounts than isochronous controls, and that the redifferentiation of both these tissues lagged behind that of isochronous controls by about 2 days. More significant, however, is the observation that the redifferentiation of MB or LB blastema in DP implants was not delayed, in contrast to the grafted blastemas of normal intercalary regenerates, nor did implants already undergoing redifferentiation dedifferentiate in most cases. There was scattered necrosis in the blastema for the first 2 days after implantation, but no necrotic cells were apparent by 4 days post-operation. In only one case (a DD implant) were nerve fibres (two)of the fin found to penetrate the blastema. Axons of the implants became highly degenerate within 2-4 days post-operation.

No dedifferentiation of stump cartilage was detected in DD implants, but was obvious in the majority of the humeri of DP implants, beginning at 6 days after implantation. Dedifferentiative activity was manifested by the presence of chondroclasts around the tip of the humerus, release of cartilage cells from their matrix within the humeral shaft and nuclear swelling and occasional mitoses in these cells. However, blastema cells never accumulated between the wrist blastema and upper arm stump, as they do during intercalary regeneration in situ.

Intercalary Regeneration in the Absence of Nerves and in the Presence of Epidermis

Epidermis-free MB-N stage regenerates were melded to upper arm stumps in fin tunnels as described in the previous experiment, except that the blastema was allowed to protrude from the tunnel so that it was re-covered with fin epidermis. Controls were as described in the previous experiment. Some specimens were fixed in Bouin's at 2-day intervals up to 14 days post-operation, serial sectioned at 10 μm and stained by the Holmes silver nitrate method. The remaining specimens were stained *in toto* for cartilage after developing for a suitable period.

Table 10.1 summarises the results from the cartilage-stained cases. Redifferentiation took place in all but one of the isolated blastemas, but the blastema often dedifferentiated and resorbed

Table 10.1
Skeletal parts regenerated by fin implants in the presence of epidermis and absence of nerves, after 16 to 28 days of development

Type of implant	*Blastema stage*	*No. of cases resorbed*	*No. of cases redifferentiated*		*% of cases regenerating the following skeletal elements*				
					H	*R-U*	*C*	*MC*	*P*
B	ER	0	1		-		100	100	100
	N	1	9		-		100	100	88. 9
		-	-		-		-	-	-
		Totals 1	10	Averages			100	100	90.9
		-	-		-		-	-	-
DƆ	MB	1	1		-		100	100	100
	ER	1	1		-		100	100	100
	N	0	4		-		100	100	75
		Totals 4	6	Averages			100	100	83.3
		-	-		-		-	-	-
DP	MB	4	9		44.4	0	100	100	77.7
	LB	3	0		-		-	-	-
	N	0	1		0	0	100	100	100
		Tatals 7	10	Averages	40	0	100	100	80

in the DD and DP implants. In nearly all cases where the blastema redifferentiated, a qualitatively complete PD array of skeletal elements was formed, due to contact of the blastema with fin epidermis. The most striking result again was the intercalary regeneration of the distal portion of the humerus in DP implants, but not of the radius and ulna.

Histological examination of 28 DP implants again revealed scattered necrosis in the blastema up to 2 days post-operation. In the majority of those cases sectioned 6 days post-operation or later, the blastema had resorbed. Humeral dedifferentiation took place in these cases, but the distal humerus did not regenerate. In the sectioned cases showing no resorption at the time of fixation, blastema redifferentiation was again found to be undelayed over the first 8 days post-operation, in contrast to normal intercalary regeneration.

Since grafted blastemas projecting from the dorsal fin do receive limited innervation from fin sensory nerves, a variation of this experiment was done in which total denervation could be temporarily accomplished. Wrist regenerates from MB-N stages were grafted with epidermis intact to upper arm stumps, or replaced at their original level. The limbs were then denervated at the level of the shoulder by transecting spinal nerves III, IV and V. Seventy-seven cases, were done, of which 32 were DD and 45 DP. All were fixed in Bouin's at 2-day intervals for up to 18 days after the operation, serially sectioned at 10 μm and stained by the Holmes silver nitrate method. Examination of the slides for the degeneration and regeneration of axons revealed that the limbs remained denervated for a minimum of 6 days.

The results of this experiment were similar to those of the corresponding fin implant experiment. Fourteen of the 32 DD cases regenerated hand without resorption of either blastema or stump. The blastemas of the remaining 18 DD cases had resorbed totally or were partially resorbed at the time of their fixation 4-18 days post-operation, and dedifferentiation and resorption sometimes extended nearly to the elbow of the stump.

The blastema redifferentiated without resorbing in only nine of the 45 DP cases. In two of these cases, fixed 16 days post-operation, intercalary regeneration of the humerus and radius and ulna had taken place and in four other cases, fixed 8 days post-operation, intercalary regeneration had begun, as evidenced by an accumulation of blastema cells between the stump and the grafted blastema. Since sequential observations of these six specimens showed that the grafted blastemas were delayed in their redifferentiation, and nerves had regenerated into all at the time of fixation, it can be assumed that they represent cases of normal intercalary regeneration. In the other three cases, fixed at 7, 14 and 21 days, respectively, no intercalary regeneration of radius and ulna took place, although one case exhibited humeral regeneration. Sequential observations of these cases indicated that they redifferentiated on a normal schedule.

To summarise the results so far, it would appear that, although stump dedifferentiation and humeral regeneration are not prevented, intercalary regeneration of the radius and ulna cannot occur under conditions that do not allow the accumulation of blastema cells before redifferentiation of the grafted blastema sets in.

Intercalary Regeneration in the Absence of Epidermis and the Presence of Nerves

In this experiment it was attempted to deprive proximally-grafted wrist blastemas of their epidermis, but not their innervation. None of the several methods tried were able to answer satisfactorily the question of whether the epidermis exerts a direct effect on dedifferentiated stump cells during intercalary regeneration, but the results nevertheless proved interesting.

In the initial experiment, a segment of spinal cord was coimplanted into a dorsal fin tunnel in contact with epidermis-free regenerate—stump melds, since Thornton (1956) had reported that limbs deplanted to the dorsal fin along with spinal cord segments became innervated by intracentral fibres and were able to regenerate normally. However, it was observed that the regenerates redifferentiated long before the implants were re-innervated by the spinal cord segment, and this approach was abandoned.

In a second experiment, epidermis-free MB-N wrist blastemas were grafted to the upper arm stump, and the regenerate plus the tip of the stump was inserted under a flap of full-thickness flank skin or covered with an envelope of leg or back skin (six cases). The technique, similar to that employed by Polezhaev and Faworina (1935), was adopted because proximally grafted blastemas become re-innervated within 2 days after transplantation, and because the normal nerve complement would be avaiable. The results were the same whether the regenerate was covered with a skin envelope, or inserted under flank skin.

Nine of the total of 15 cases were fixed 20-28 days post-operation and stained *in toto* for cartilage; the remainder were fixed in Bouin's 6-10 days post-operation, serial sectioned at 10 µm, and stained with silver nitrate or iron haematoxylin and eosin. In four of the nine cartilage-stained specimens, the skin covering the blastema became translucent a few days post-operation, and normal limbs regenerated. The sectioned cases, most of which appeared to be regenerating normally at the time of fixation, showed that the dermis had dis-appeared in the translucent area, thus allowing epidermis to contact the underlying tissue. However, regeneration in the four cartilage-stained cases mentioned could have been either by intercalary regeneration or by distal regeneration from the stump after resorption of the blastema. A normal intercalary regenerate has two 'zones' of cells 6-8 days post-operation, a dense distal zone representing the graft and a less dense proximal zone of cells apparently derived from the stump. The regenerates of

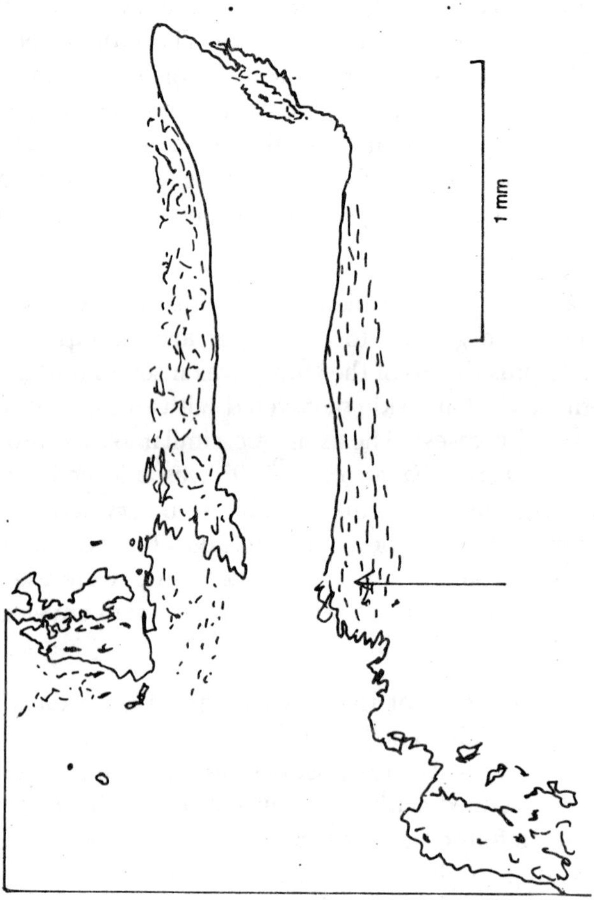

Fig. 10.3. Notch stage wrist regenerate grafted to the proximal third of an upper arm stump, 28 days post-operation. The wrist regenerate resorbed, but a relatively normal humerus regenerated. Arrow indicates the level at which the graft was placed.

the six sectioned cases, in the present experiment did not show this zonation, but looked like normal distal regenerates. Thus, in the four cases showing complete regeneration, it is probable that the blastema resorbed followed by regeneration from the stump. Why the balstema should resorb under the conditions of this experiment in puzzling, especially since normal intercalary regeneration takes place when EDTA-stripped wrist regenerates are grafted to upper arm stumps and are re-covered with epidermis migrating from the upper arm skin (Stocum, unpublished experiments).

Of the five remaining cartilage-stained specimens, the blastema had resorbed in four. In three of these four cases, the humerus regenerated, but a complete limb was not produced by distal regeneration. In the other case of this group the humerus also regenerated, but a late bud regenerate was present at an angle to its distal end at the time of fixation (20 days post-operation). These five cases show that an amputated limb segment can complete itself under the conditions of the experiment, and that distal regeneration after blastemal resorption requires input from a functional epidermis, but do not tell us anything about epidermal requirements for intercalary regeneration.

In the single remaining cartilage-stained case, the blastema mesenchyme did not resorb, but redifferentiated metacarpals and phalanges after coming into contact distally with epidermis 6 days post-operation. However, even though all the disposing factors appeared to be present, intercalary regeneration of the radius and ulna did not take place, although the humerus regenerated. It is possible that this is another case where rapid blastema redifferentiation took place, preventing radius and ulna formation.

In a third experiment, comprising 16 cases, fully-differentiated hands were grafted to upper arm stumps. Eleven of these were stained *in toto* for cartilage, and five were fixed in Bouin's 8-12 days post-operation, serial-sectioned at 10 µm and stained with iron haematoxylin and eosin. All of the cartilage-stained specimens regenerated the humerus, but not the radius and ulna. The humeri of these cases and the cases described in the immediately preceding experiment were all much more normal in size and form than those regenerating

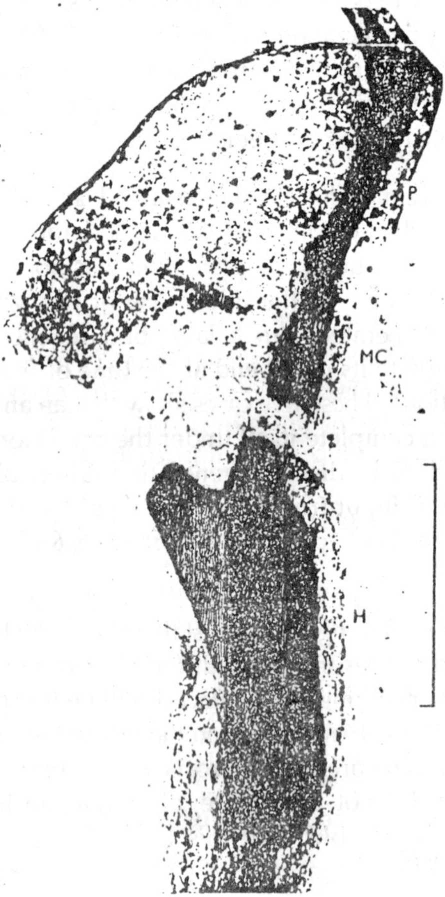

Fig. 10.4. Notch stage wrist regenerate stripped of epidermis with EDTA, grafted to the mid-upper arm stump and re-covered with an envelope of back skin, 27 days post-operation. Contact was made between the mesenchyme and the distal epidermis and the graft grew out and redifferentiated metacarpals (MC) and phalanges (P). The humerus (H) completed itself distally, but there was no regeneration of radius and ulna.

under nerveless conditions. The sectioned cases showed, however, that they were derived from an initial population of dedifferentiated stump cells that appeared no larger than that seen in the absence of nerves. There was no longer accumulation of blastema cells before humeral redifferentiation set in, as there is in normal intercalary regeneration. However, once the humeral pro-cartilage condensation had formed, it exhibited a high incidence of mitotic figures. Apparently, the presence of nerves alone is enough to ensure the normal growth of the humeral pattern that is established in an initially small population of dedifferentiated cells, but the results again demonstrate that additional input involving an undifferentiated blastema component is required for establishing the pattern and or outgrowth of more distal intercalary structures.

Discussion

The essence of limb regeneration is the distal transformation of an initially small population of dedifferentiated stump cells into all of the missing limb structures. Distal transformation is necessary for both distal and intercalary regeneration and there are two major requirements for its occurrence or detection. First, the complete circumference of the limb apparently must be represented at the amputation surface. Secondly, there must be a sufficient accumulation of blastema cells. This is normally provided by mitosis of the dedifferentiated cells. Mescher and Tassava (1975), and Mescher (1976) have shown that in distal regeneration both the nerve and the epidermis are required for this cell division. They found, in experiments with adult newts, that cells of an amputated limb stump dedifferentiate and enter the cell cycle in the absence of either nerves or epidermis, but do not continue to divide and accumulate as a blastema.

The present study demonstrates that an upper arm stump joined to a wrist regenerate can dedifferentiate in the absence of nerves, epidermis, or both, but that blastema cells cannot accumulate and no intercalary regeneration of the radius and ulna can take place under these conditions. Nevertheless, under these same conditions, the humerus is able to regenerate. This interesting and potentially important result suggests that the initial small population of dedifferentiated stump cells is able to undergo autonomous distal transformation to complete the segment of its origin, but that neural and blastemal influences are required for the production of more distal structures. S. Bryant has also observed intercalary regeneration of the humerus after grafting hands to upper arm stumps.

From the results reported here, two major factors appear to be necessary for the further proliferation and distal transformation of blastema cells leading to the intercalary regeneration of the radius and ulna. First, innervation is required, for a proximally grafted blastema covered with epidermis most often dedifferentiates and resorbs in the absence of nerves. With respect to innervation, then, the requirement is the same for intercalary regeneration as for distal regeneration. Secondly, the regenerate that is grafted proximally must be *undifferentiated.* This is shown by the following facts : (1) lack of radius-ulna formation was almost always associated with the normal temporal course of redifferentiation of the regenerate instead of the characteristic 8-day delay in redifferentiation; and (2) the radius and ulna failed to regenerate even in the presence of a full complement of nerves when a differentiated hand was grafted to the upper arm.

The role of the undifferentiated blastema in the distal transformation and accumulation of blastema cells that produces a complete intercalary regenerate could be mediated either through its mesenchymal or epidermal component, or both. An undifferentiated epidermis appears necessary for the production or maintenance of the undifferentiated state of a grafted regenerate during intercalary regeneration, because regenerates melded to upper arm stumps in the absence of epidermis and nerves almost always redifferentiate without delay, while the majority of those that are re-covered with fresh epidermis in the absence of innervation dedifferentiate and resorb. Therefore, it is logical to suspect that the epidermis is at least an indirect factor in the distal transformation leading to radius and ulna formation. However, a direct epidermal influence on dedifferentiated stump cells during intercalary regeneration was not established by the present experiments, due to failure of the grafted blastema to survive under the techniques employed. And, this type of experiment may not be able to test for a direct epidermal effect, for if an undifferentiated blastema is a requirement for complete intercalary regeneration, and epidermis is necessary to produce or maintain this condition, we have no way to distinguish between the epidermis and the mesenchyme as the necessary component.

Whatever the eventual answer to the question of how the epidermis is required for intercalary regeneration, its direct influence on the mesenchymal outgrowth of distal regeneration and embryonic limb development as well-documented. Therefore, I would like to discuss how the epidermis and the nerves might work together to promote regeneration.

From studies on the dedifferentiative phase of limb regeneration, Tassava & Mescher (1975) have proposed that the epidermis keeps dedifferentiated cells in the cell cycle and prevents their immediate redifferentiation, while the nerves are required for the cells to traverse the G_2 portion of the cycle. Inhibition of the function of either nerve of epidermis would thus lead to failure of blastema formation. This is an interesting hypothesis, although there may be no need to assume different functions for the nerve and epidermis, as discussed below.

Nerves are not necessary for avian or amphibian embryonic limb bud development nor the morphogenesis of established limb regeneration blastemas but the limb ectoderm or epidermis is necessary for the continued outgrowth and morphogenesis of these structures presumably because of its proposed liability effect on the subjacent mesenchyme. Aneurogenic larval urodele limbs are also able to regenerate. To provide a unified framework for these observations, Singer (1965) proposed that non-neural cells of the amphibian limb bud produce their own mitosis-promoting factor, but that its synthesis is suppressed and supplanted by ingrowing nerves during development. The limb therefore becomes nerve-dependent for regeneration unless it has remained aneurogenic. The factor must be present at a certain threshold concentration to be effective.

This elegant concept provides a simple quantitative explanation for the observed differences in the neural requirement for limb development and regeneration, and does not require the nerve to have one function and the epidermis another. The idea would state that nerves and ectoderm or epidermis produce the same factor which has the effect of maintaining a subjacent zone of undifferentiated, dividing mesenchyme cells. The ectoderm would produce enough of the factor for initial limb development, but ingrowing nerves would supplant this function and adjust downward the amount produced by the epidermis so that regeneration comes to depend on both. In the continued absence of nerves (aneurogenic limbs) this adjustment would not take place and the epidermis would continue to produce the factor at its original rate. The factor might have a rather limited life which would restrict the diffusible range of the epidermal influence to an apical region; the neural influence, however, would be distributed throughout the blastema. In both innervated and aneurogenic regenerating limbs, cells would remain as part of the dividing and undifferentiated population until outgrowth left them proximal to the range of the AER contribution. These cells would then be below the threshold concentration of factor, and would therefore redifferentiate. Redifferentiation would thus proceed in the observed PD sequence. These ideas are consistent with the fact that aneurogenic limbs fail to regenerate when covered with skin from innervated limbs, but do so if their internal tissues are replaced by those from an innervated limb and also fit the observation that originally aneurogenic limbs can be oscillated between regenerative nerve dependence and independence when transplanted to an innervated host.

However, whether the epidermis acts to promote the outgrowth of limb buds and regenerates in the inductive sense referred to above is still controversial. Amprino has justifiably argued that limb mesenchymal outgrowth may not be inductively dependent on ectoderm or epidermis. The mesenchyme may be protected by and its pattern moulded to some extent by the epithelial covering, but outgrowth may be generated in large part by the activities of the mesenchyme itself. This controversy revolves around whether the effect of ectodermal or epidermal deletion is to remove an outgrowth stimulus, or to expose the mesenchymal cells to adverse conditions leading to necrosis of cells that would normally form terminal limb parts, and has not been resolved.

The major issue of pattern formation along the PD axis of the regenerating limb is the mechanism by which distal transformation occurs and the relationship between this process and the interactions of the tissues influencing the outgrowth of the blastema. The nerve is unlikely to have other than a mitosis-promoting function since a variety of nerve sources and configurations are effective in regeneration. Furthermore, it has been shown that denervation affects the protein composition of the blastema much more severely at late regenerative stages when the blastema is independent of the nerve for its morphogenesis, suggesting that the nerves do not affect morphogenetic processes *per se*. The function of the epidermis in pattern formation, however, could be more complex than the outgrowth-promoting role that has been postulated for it.

There are several current models of PD pattern formation that are applicable to limb regeneration in which the epidermis could have a role other than in outgrowth *per se*. In the simplest of these models, that of sequential and autonomous pattern specification, the zone of labile mesenchyme cells constitute a 'progress zone' in which distal transformation takes place. The longer the time spent by cells in the progress zone, the more distal the pattern they will form. It is important to realise that distal transformation in this model can be either directly proportional to numbers of cell divisions in the progress zone, in which case the epidermis might have only the function of keeping the cells in the cell cycle, or it could be a function of an ageing process controlled by production of an epidermal factor in which case the-epidermis would be functioning in a dual capacity.

The synchronous specification model postulates a gradient of positional information extending along the PD axis of the blastema which specifies all the parts of the pattern at once. Here the epidermis could be involved in the production of the positional information gradient, as well as having an outgrowth function. Although there is evidence from studies on the avian limb bud that the AER does not act as a level-specific inducer of pattern it could act as the source or sink of a positional information gradient.

In the sequential interaction model the spatial pattern of differentiation is imposed upon cells as they pass out of range of the epidermal outgrowth factor by interaction with previously specified proximal parts of the pattern. The epidermis has only an outgrowth promoting influence in this model. It has already been observed that blastema cells might retain a 'memory' of their level of derivation and that this memory could be used to initiate a PD sequence of pattern-forming events in the undifferentiated mesenchyme. The fact that intercalary regeneration of the humerus appears to be an autonomous process is consistent with this idea. The memory concept is also compatible with the other two models of pattern formation, where it would define the proximal boundary of the regenerate.

In conclusion, it is clear that, while we have described a large number of experimental phenomena for the regenerating limb, and a plethora of models have been proposed, we do not yet have a good idea of how distal transformation occurs to regenerate a limb. A great deal of work remains to be done at the tissue and organ level of biological organisation before the relationships between outgrowth, specification of spatial patterns of redifferentiation and the tissue interactions which control them during regeneration, are understood.

11

Regeneration of the Axolotl Limb

The development in *Ambystoma* is indirect. Its larva is the *axolotl* and wherever it develops in the absence of iodine it becomes sexually value and starts reproducing sexually. This breeding behaviour is called ncotenry. These neotenic axolotls spend all of their normal life as aquatic larvae. As is the case with all larval urodele amphibians, axolotls are capable of regenerating their appendages remarkably well. In recent years, they have become increasingly important as laboratory animals because they are easy to maintain and breed readily over the winter months. The limbs of the larvae are particularly well suited to studies of pattern regulation because normal limbs faithfully replace the limb parts distal to the level of amputation in an ordered histological sequence, and the skeleton, muscles, and skin provide clear anatomical markers for the results of tissue level manipulations.

In this discussion, a theoretical framework for the control of pattern regulation will be analyzed in terms of a series of recent experiments. The intention of the discussion is twofold; the experiments to be outlined will reveal some of what we know of the cellular interactions occurring during distal outgrowth and will indicate how a theoretical framework for these interactions molds the design of experiments and evolves as new data concerning specific tenets of the model become available. Some of the basic features of the polar co-ordinate model have remained unscathed during the last six years, whereas others have been discarded or have been refocused.

Prior to describing the series of experiments that will form the bulk of the discussion, the model itself will be outlined.

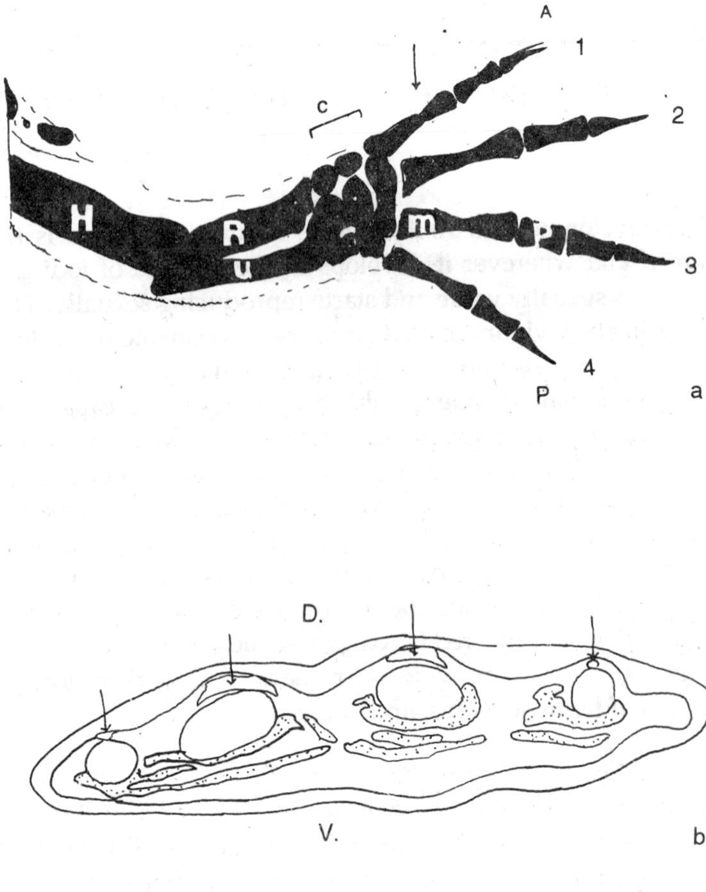

Fig. 11.1. The axolotl forelimb bears numerous clear structural markers that can be used to analyse the results of tissue level grafting experiments. (a) The normal skeleton (magnification x 15) of the forelimb demonstrated by Victoria blue wholemount staining. Several skeletal markers are evident. For example,

the third digit has three phalanges, digits three and four articulate with individual carpals and the ulna overlaps the humerus at the elbow joint. (b) A camera lucida drawing of a 10-micron wax section cut at the proximal metacarpal level (arrow on a) revealing the dorsal to ventral differences in the muscle pattern (magnification x 30). The ventral muscles (dotted regions) are a complex arrangement of four muscles which are virtually continuous across the anterior to posterior extent of the limb. In contrast, the dorsal muscles (arrows) are discrete semicircular shaped masses covering the dorsal side of the metacarpals (see 14-17 for a more detailed description of the muscles). H, humerus; R, radius; U, ulna; c, carpals; m, metacarpals; p, phalanges. The digits are numbers 1 to 4 from anterior (A) to posterior (P). On the section, D is dorsal, and V is ventral.

The Polar Co-ordinate Model

The polar co-ordinate model is a formal representation of a two-dimensional co-ordinate array of positional values with a circumferential and a radial component. The model was devised in an attempt to explain how pattern regulation was achieved in structures which regenerated by the addition of new cells, as opposed to such animals as *Hydra,* which can regenerate in the absence of cell division by reorganizing the tissue which remains. Regeneration by the addition of new cells occurs in such animals as the cockroach and the fruit fly when tissue damage is incurred. This type of pattern regulation is termed *epimorphosis.* In its initial form, two basic rules were suggested to explain the epimorphic regulation of pattern when normally non-adjacent cells were confronted following surgical intervention. The first of these rules stated that a complete circumference of positional values needed to be present before outgrowth could commence. This complete circle rule was initially put forward in order to explain the appearance of supernumerary limbs that appear after axial misalignment of blastemas and stumps. In a sense, the rule was in conflict with the remainder of the model because it relied on a global realization within the limb that a complete asymmetrical circle was present before outgrowth could begin. The remainder of the model relies solely on local cell-cell contacts and the intercalation of new cells by cell division that restore circumferential or radial continuity after the confrontation of of normally non-adjacent cells. The polarity of restoration of continuity by intercalation is controlled by the second rule which states that this always occurs by the shortest possible route. In this way, the actual positional values intercalated should always be predicted except when maximum points of discontinuity are confronted when a choice of two paths of intercalation are evident.

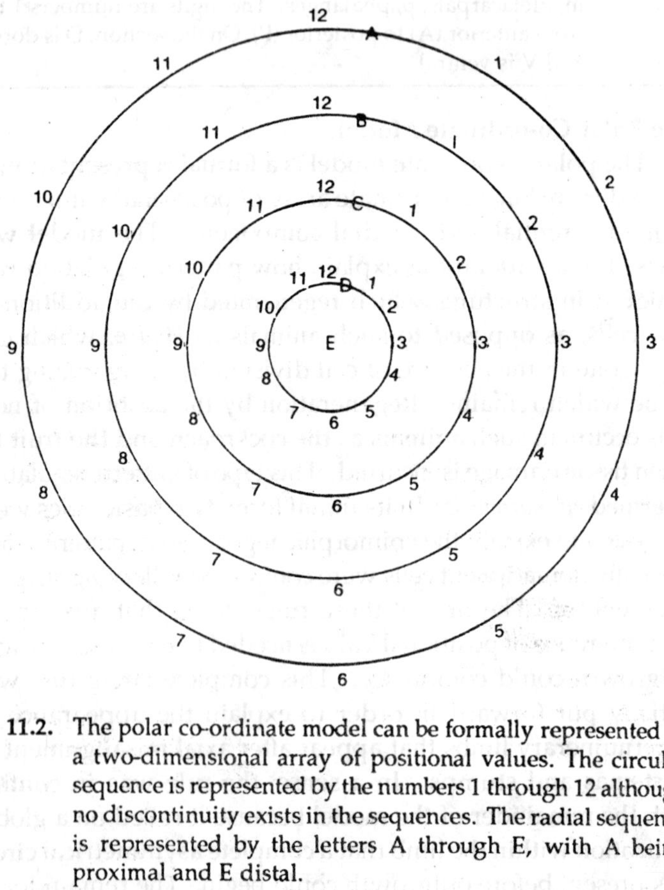

Fig. 11.2. The polar co-ordinate model can be formally represented as a two-dimensional array of positional values. The circular sequence is represented by the numbers 1 through 12 although no discontinuity exists in the sequences. The radial sequence is represented by the letters A through E, with A being proximal and E distal.

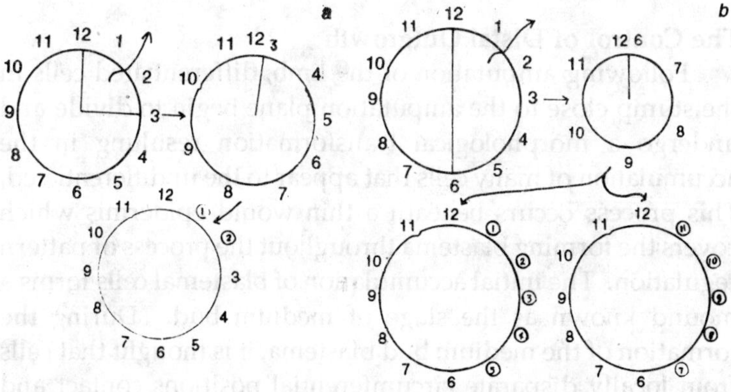

Fig. 11.3. Diagrammatic representation of the shortest route intercalation rule that governs the positional values held by new cells produced following contact of cells from normally non-adjacent positions. (a) After removal of a quadrant bearing only two positional values, intercalation restores the original pattern following the healing of the cells at the cut edges. (B) If half of the circumferential set is removed two possible paths can result. In one case, regeneration of the original pattern will occur. In the other, a symmetrical pattern is created following the duplication of the values already present. In all drawings, the newly intercalated positional values are ringed.

Soon after the model first appeared, numerous experimental results demonstrated that the complete circle rule was not required to explain pattern regulation. It is now clear that tissues bearing incomplete asymmetrical circles or symmetrical circles can commence outgrowth. This is evident from experiments concerned with both the outgrowth of stumps bearing symmetrical tissues and from the structure of supernumerary limbs formed following ipsilateral blastemal rotations.

In the first section of experimental results presented, the likely controls of normal outgrowth will be discussed. The possible interactions leading to the formation of supernumerary limbs will be outlined later on.

The Control of Distal Outgrowth

Following amputation of the limb, differentiated cells in the stump close to the amputation plane begin to divide and undergo a morphological transformation resulting in the accumulation of many cells that appear to the undifferentiated. This process occurs beneath a thin would epidermis which covers the forming blastema throughout the process of pattern regulation. The initial accumulation of blastemal cells forms a mound known as the stage of medium bud. During the formation of the medium bud blastema, it is thought that cells from locally disparate circumferential positions contact and intercalate cells with intervening positional values. These local interactions occur at points all around the circumference and the new cells produced take on the next most distal radial value to ensure that new pattern elements are produced only in a distal direction. The continuous growth of the blastema coupled with local contacts between cells in this way at the tip of the blastema replaces the lost pattern in a proximal to distal sequence. This simplistic way of thinking about how normal distal outgrowth may occur gains support from experiments in which symmetrical limb stumps are surgically created and amputated. From the results of these experiments, it is now clear that the amount of distal outgrowth involving the sequential production of new radial levels is intimately linked to both the nature of the circumferential sequence at the amputation plane, and the mode of healing that controls which circumferential cells actually make contact during outgrowth.

Symmetrical limb stumps have been created surgically in both the upper and lower limb regions. The basic experiment involves cutting one half of the limb out from upper or lower limb positions and exchanging tissues from contralateral limbs. Double anterior and double posterior constructions have been made in most cases, although double dorsal and double ventral constructions have also been examined. The amputation of such limbs produces a variation of structures ranging from extensive symmetrical regenerates to nothing at all.

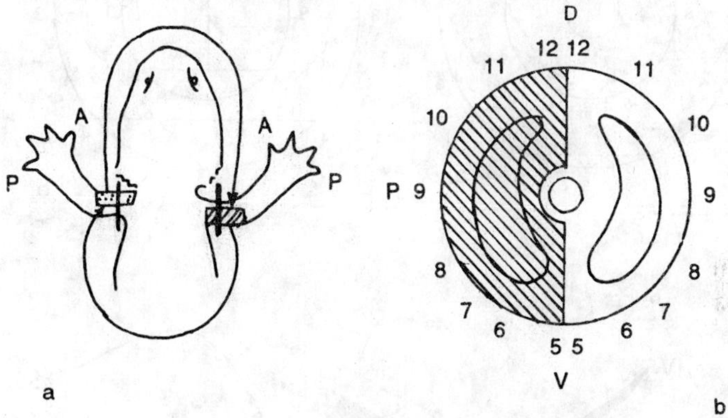

Fig. 11.4. (a) Diagrammatic representation of the operation performed to create double anterior and double posterior symmetrical stumps. (b) An end-on view of the amputation plane of a double posterior stump. The graft is hatched and the theoretical distribution of circumferential positional values is biased so that the posterior half of the limb bears greater than half.

The very fact that any regeneration is achieved is evidence that the complete circle rule is not upheld. Several other directives emerge from an analysis of the data from these experiments. It is clear from operations performed on the upper forelimb that double posterior constructions consistently regenerate more structure than double anterior constructions. Furthermore, the degree of distal outgrowth is dependent upon the length of time that the initial graft is allowed to heal prior to amputation. The longer the graft healing time, the less the regeneration that is seen following amputation. This effect has also been detected in similar experiments in the thigh region of the axolotl leg. Two further conclusions have been drawn from these results. The reason for posterior tissues being more productive than anterior tissues in symmetrical combinations is assumed to be due to there being greater than

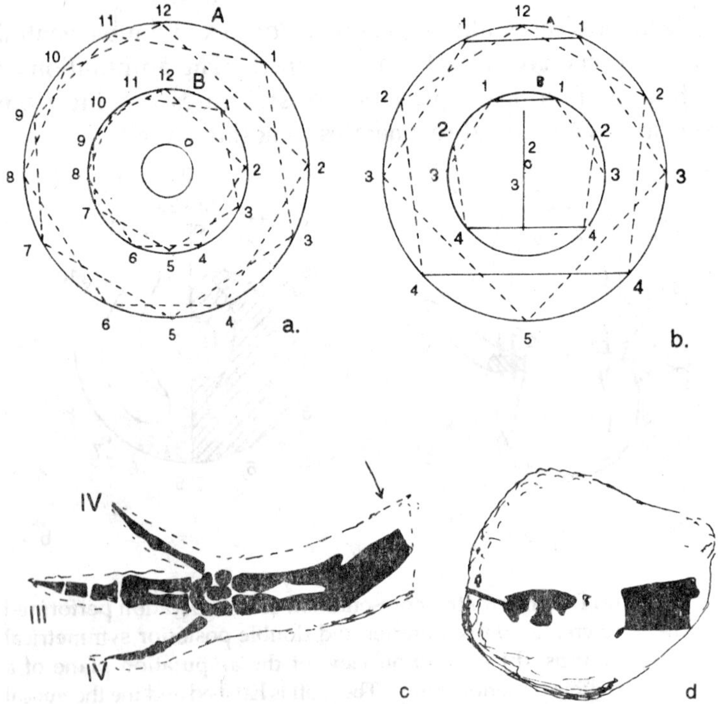

Fig. 11.5. Representation of the theory of short arc intercalation. (a) Diagrammatic view of normal distal outgrowth following amputation at radial level A. Cells are thought to contact each other over short circumferential distances during initial blastema accumulation (dotted lines). Contacts between normally locally non-adjacent cells leads to intercalation of intervening cells whose new positional values are determined by the shortest route intercalation rule. (b) Short arc intercalation in a symmetrical, double-anterior upper arm stump in which the graft has healed for a month before amputation. The symmetrical arrangement of circumferential values and the allowance of non-productive cell contacts leads to rapid loss of positional values and the cessation of regeneration. (c) Victoria blue stained whole mount of a double anterior limb amputated in the upper arm after one month of healing. This limb failed to regenerate any clear structure (magnification

x 16). The arrow points to the level of amputation. (d) A double posterior, symmetrical limb regenerated following amputation of a double posterior upper arm in which the graft had healed for only five days (magnification x 18). In contrast to long healing time experiments, regeneration continues to a distal limb region and digits are formed. In such cases, the wound between the graft and host stump tissues is thought to prevent non-productive cell contacts. The leader points to the regenerated cartilage.

half the number of circumferential positional values on the posterior side of the upper limb. The role of tissue healing is thought to involve the structural constraint of contacts between cells on either side of the wound created during surgery.

This last point can be most clearly thought of in terms of the simple model discussed above for normal limb regrowth. It is thought that after long periods of healing, cells on the two sides of the wound can interact freely. In this case, cells will interact locally as in normal outgrowth and, because of the symmetry in the experimental cases, this will result in a number of cell contracts occurring between cells with the same or normally adjacent circumferential value. This type of contact will not stimulate intercalation. In addition, no local contact around the circle will result in the intercalation of the mid-line circumferential values at the next most distal level to that of amputation. The outcome of these two properties of symmetrical circles will be a tapered symmetrical outgrowth outgrowth that forms from a blastema containing few cells. In contrast to this effect, which occurs after long periods of healing, short healing grafts regenerate far more structure. This is assumed to be due to the prevention of contacts between cells on either side of the central wound, and consequently the prevention of non-productive cell interactions. The result of this will be more cells available for intercalation at the next most distal level. The result of amputating such short healing time limbs should be a distally tapering structure that regenerates to a more distal pattern level than the longer healing type of graft.

In many cases, regenerates of this class distally transform to the digits.

Thus far we have examined only symmetrical outgrowths from upper arm stumps. Two further experiments have been performed on axolotls which examine the regenerative ability of symmetrical double posterior and double anterior lower arms. The first of these experiments clearly suggested that lower arm symmetrical constructs behave differently from comparable upper arm constructions, in two ways. First, no clear healing time effect is seen in the axolotl lower arm. Amputations of symmetrical stumps immediately after surgery produces similar degrees of outgrowth of those amputated after a month of healing. Secondly, the degree of regeneration of double posteriors and double anteriors is comparable. In the second experiment, double posterior limbs that had regenerated from initial short healing time upper arm amputations of symmetrical double posterior constructions produce expanded patterns when amputated in the lower arm. These types of results strongly suggested that the lower arm was behaving differently from the upper arm in some fundamental way. Not only is the distribution of values around the circumference thought to be more even, but the constraints on contacts between the cells around the circumference seem to be different.

This point can be clearly demonstrated by examining how distal expansion of a double posterior limb amputated in the lower arm can occur. In the majority of cases, double posterior upper arms amputated immediately make symmetrical limbs bearing between three and six digits. When reamputated in the lower arm, limbs bearing up to seven digits regenerate with a normally anterior digit appearing in the middle of the pattern. In terms of shortest route intercalation, this can only occur if cells with the dorsal and ventral extremes of the circumferential set of values interact. The question at hand is why should this occur in the lower arm but not in the supper arm ? The simplest conclusion, if local cell-cell contacts are

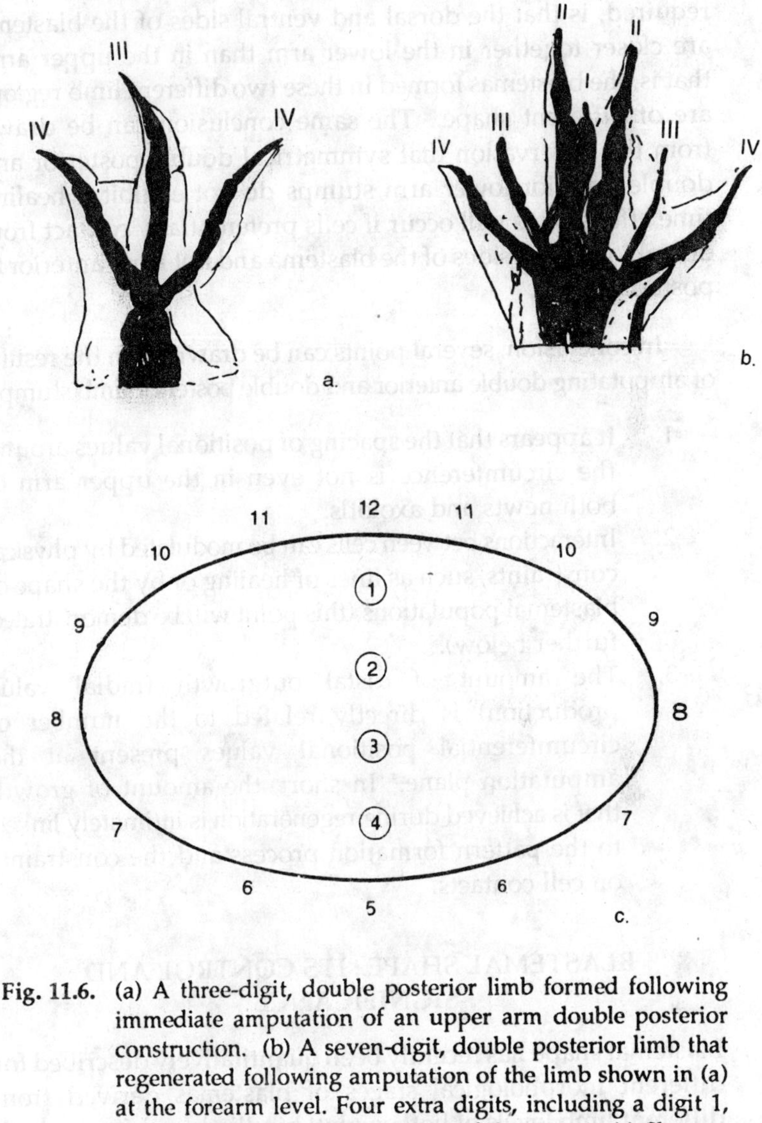

Fig. 11.6. (a) A three-digit, double posterior limb formed following immediate amputation of an upper arm double posterior construction. (b) A seven-digit, double posterior limb that regenerated following amputation of the limb shown in (a) at the forearm level. Four extra digits, including a digit 1, have been added in the midline of the pattern (c) Shortest route intercalation predicts such an event only if local contact between cells on the dorsal and ventral sides of the blastema can occur. Intercalated positional values are ringed.

required, is that the dorsal and ventral sides of the blastema are closer together in the lower arm than in the upper arm; that is, the blastemas formed in these two different limb regions are of different shape. The same conclusion can be drawn from the observation that symmetrical double posterior and double anterior lower arm stumps do not exhibit a healing time effect. This will occur if cells preferentially contact from dorsal to ventral sides of the blastema and not from anterior to posterior sides.

In conclusion, several points can be drawn from the results of amputating double anterior and double posterior limb stumps.

1. It appears that the spacing of positional values around the circumference is not even in the upper arm of both newts and axolotls.
2. Interactions between cells can be modulated by physical constraints, such as lines of healing or by the shape of blastemal populations (this point will be demonstrated further below).
3. The amount of distal outgrowth (radial value production) is directly related to the number of circumferential positional values present at the amputation plane. In short, the amount of growth that is achieved during regeneration is intimately linked to the pattern formation process and the constraints on cell contacts.

BLASTEMAL SHAPE : ITS CONTROL AND SIGNIFICANCE

Blastemal shape has recently been quantitatively described for different morphological stages of blastemas derived from different limb levels of both axolotl hindlimb and forelimb. It has clearly been demonstrated that upper arm and leg blastemas are round in cross-section whereas lower arm and leg blastemas are elliptical. This observation is entirely consistent with the predictions made from the results of the experiments just discussed.

This analysis of blastemal shape was based upon a simple morphometric analysis. Limbs were amputated at the required level and allowed to regenerate to the stage of medium bud, medium to late bud, and palette. At the required stage, limbs were fixed, wax embedded, and serially sectioned. The distance of the main anterior to posterior and dorsal to ventral axes were measured at 100-μ levels from the distal tip and a simple ratio of these values calculated. This ratio, r, was then plotted for each blastemal stage at each level of amputation.

It is clear from these graphs that the lower limb blastemas are significantly more elliptical than the round upper limb blastemas. It should be emphasized that this shape analysis was carried out subsequent to the surgical symmetrical limb stump experiments and indicate therefore that structural constraints can be predicted from such an approach. At the same time, however, blastemal shape may be used to examine the problem of pattern formation. If the control of shape can be established, then it should be possible to alter blastemal shape and the critical relationship between shape of tissue fields and pattern formation can be further examined. For this reason, a further set of experiments were performed in an attempt to elucidate the tissue controls of blastemal shape.

At the outset of these experiments, the simple questions to be examined involved the role of the stump in the control of blastemal shape. Clearly the stump is of maximum importance during the early phases of regeneration because all blastemal cells derive from it. However, it is already known that by the stage of MB the pattern regulation mechanisms of the regenerate are autonomously controlled by the blastema and are not influenced by the stump from which it derives. We wished to know if the same was true of the control of blastemal shape. An indication that this was the case came from observing a change in shape from round to elliptical during outgrowth of a palette staged upper arm blastema. In this case, the flattening was occurring spontaneously within the blastema because it was forming on an upper arm stump. Further evidence for an autonomous blastemal control came from surgical experiments where specific tissues of the stump were removed prior to amputation to see if an abnormal stump would produce misshapen blastemas.

In these experiments, individual fore and hindlimbs had either the cartilage, muscle, or dermis removed. Removal of one bone, either the radius or ulna, from the lower forelimb and removal of stump musculature from the thigh region of the leg had little effect upon blastemal shape as compared with sham operated controls. These results suggested that the role of the stump in governing blastemal shape involved cryptic, level specific information rather than a purely mechanical role. This conclusion was supported by the removal of the dermis from the stump which caused a flattening of the blastemas derived from both the upper and lower limb levels.

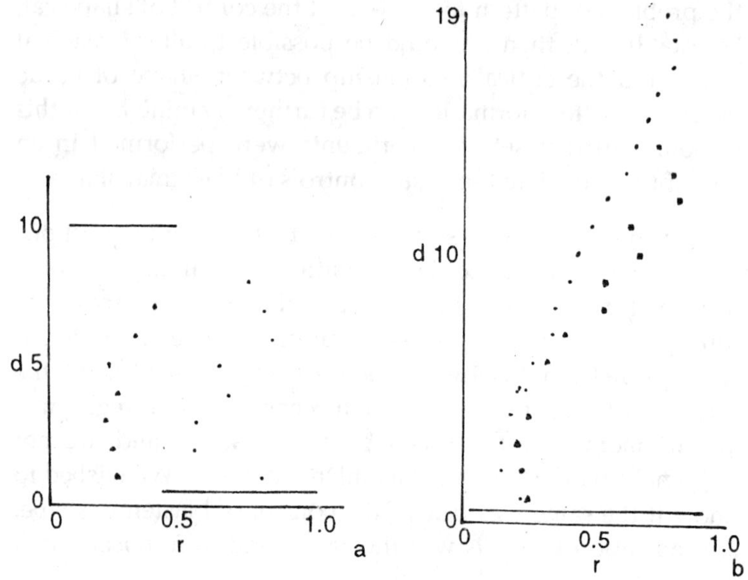

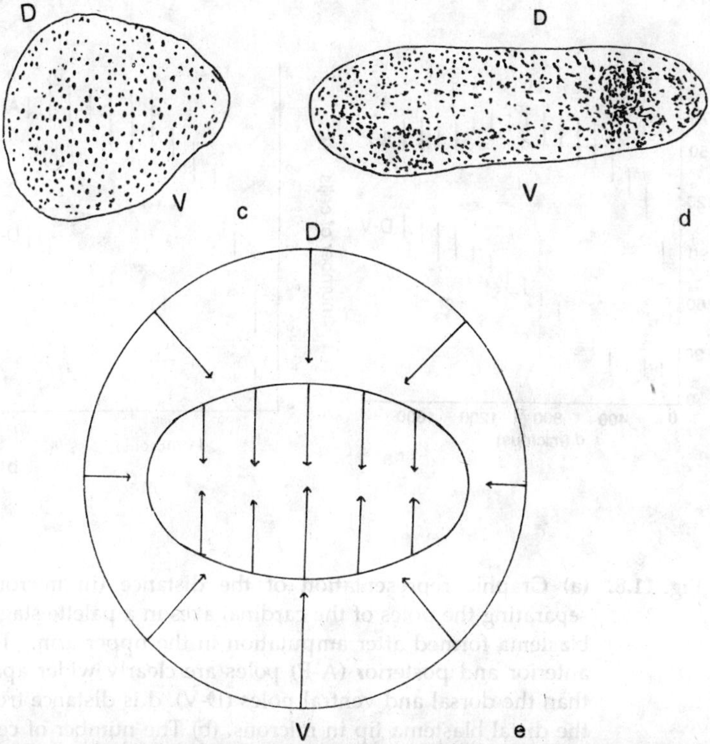

Fig. 11.7. The shape of blastemas formed at different proximal to distal limb levels and at different stages. r = shape ratio; d = distance from the blastemal tip in 100 micron steps. D = dorsal; V = ventral. The arrows on the graphs represent 95% confidence limits for the range of r values for each blastema population. Each point represents the mean of between 3 and 5 separate blastemas at corresponding levels from the distal tip. (a) Upper arm (●) and lower arm (Δ) medium bud shape ratios. The lower arm MB is markedly more elliptical than the upper arm MB blastema. (b) Upper arm palette staged blastema shape ratios (●). The blastema flattens distally after being initially rounded. This point is emphasised by placing the measurement of upper (■) and lower (Δ) arm MB blastemas shown in (a) on the same graph. (c) A typical transverse section of a rounded upper arm MB blastema. (d) A typical transverse section of an elliptical lower arm MB blastema. (e) A hypothetical diagram of the healing modes in blastemas at upper and lower limb regions. In the upper arm (outer circle), the healing mode is radial. In the lower arm, the healing mode is preferentially dorsal to ventral.

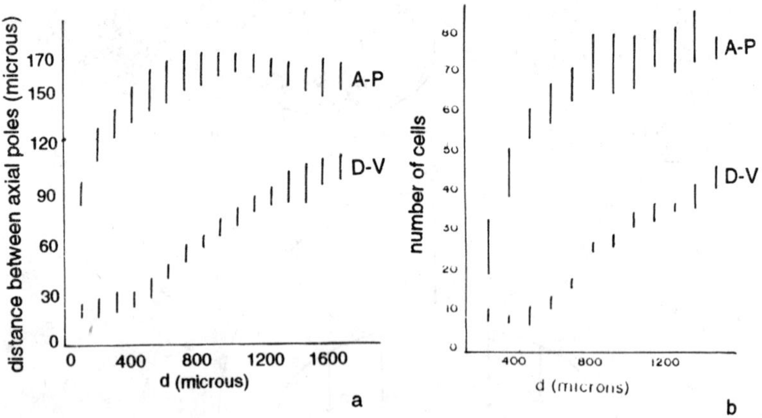

Fig. 11.8. (a) Graphic representation of the distance (in microns) separating the poles of the cardinal axes in a palette staged blastema formed after amputation in the upper arm. The anterior and posterior (A-P) poles are clearly wider apart than the dorsal and ventral poles (D-V). d is distance from the distal blastema tip in microns. (b) The number of cells separating the cardinal axes in a palette staged blastema formed after amputation in the upper arm. In both graphs the points represent the means of 5 measurements and the bars represent standard errors.

The alterations in shape of blastemas from round to elliptical is consistent with the mode of healing and interaction that is predicted from the analysis of regeneration of symmetrical limb stumps. With regard to this particular point, an analysis of the number of cells separating the dorsal-ventral and anterior-posterior poles of blastemas and measurement of the dimensions of blastemas with reference to these axes demonstrates that these facets of blastema outgrowth are entirely consistent with the process of short arc intercalation and the proposed constraints on healing following amputation at different limb levels. Geraudie and Singer have recently demonstrated, using the scanning electron microscope, that newt limb blastema cells have cellular processes that may be up to 50μ long. In the distal region of a palette staged blastema formed after upper arm amputation, the dimensions of and cell counts along the cardinal axes are consistent with the possibility of cells from dorsal and ventral sides of the blastema contacting directly. For example, in the distal 700m of such blastemas, which is the region forming the limb region distal to the elbow, the distance from dorsal to ventral extremes varies from 20 to 50m and are separated by between means of 7 and 26 cells. In contrast, the anterior and posterior poles are between 84 and 151m apart and are separated by between means of 25 and 67 cells. Thus the likelihood of cells contacting preferentially from dorsal to ventral sides of the blastema is higher than the likelihood of anterior and posterior cells contacting and the dimensions are consistent with a local cell contact model. Furthermore, dimensions and cell counts analyzed in upper arm MB blastemas are consistent with a radial healing mode.

Conclusions: The Importance of Moderators of Cell-Cell Contact

The data presented in this discussion are consistent with a short range cell-cell contact model controlling pattern regulation in the amphibian limb. It is clear, however, that trying to explain the results solely in terms of the formal representation of the polar co-ordinate model is inadequate and that secondary considerations, such as healing constraints and blastemal shape are needed to make the explanations more complete. This point is further emphasized by recent experimental findings where regenerating limbs bearing clear structural discontinuities have been demonstrated. These experiments will be briefly outlined as the conclusion to this discussion to demonstrate the problems that face attempts to elucidate the basis of pattern regulation in the urodeles.

The first demonstrations that limbs could form bearing structural discontinuities were made by Maden and Maden and Mustafa who analyzed the muscle patterns in supernumerary limbs produced following ipsilateral blastema rotations in the axolotl. It is now clear that these extra limbs bear a variety of dorsal-ventral pattern characteristics despite the fact that the anterior-posterior pattern appears normally asymmetric. Four basic limb anatomies have been found. These include double dorsal and double ventral limbs, limbs with part of their pattern symmetrical (double dorsal and double ventral) and part asymmetrical, limbs with mixed handed patterns where half of the pattern is dorso-ventrally reversed compared with the other half, and normal asymmetrical limbs.

The supernumerary limbs comprising part symmetrical and part asymmetrical patterns or mixed-handed patterns have clear structural discontinuities within them. The formal polar co-ordinate model and any other model in which clear discontinuities should be smoothed out fail to explain how such limb patterns can occur. The same is true of the recent descriptions of co-ordinate-free models based simply on rules of pattern continuity that do not discuss likely cellular mechanisms. An additional problem is that these supernumerary limbs are derived from interactions between positionally mismatched blastemas and stumps and no way is currently available for elucidating the details of these interactions. For this reason, we have recently surgically created forearms that are anatomically equivalent to the mixed-handed supernumerary limbs. Such forearms were made by splitting the axolotl limb between digits 2 and 3 and separating one complete half of the forearm to the elbow between the radius and ulna. Anterior or posterior halves were then exchanged between left and right limbs and sutured into the vacant lower arm position such that the dorsal and ventral axes of graft and host limb tissues were opposed but the anterior to posterior axes were normal. Such constructions were allowed to heal for one month before amputation in the mid forearm position or amputated immediately. In both instances, such mixed-handed limbs regenerated essentially the same mixed-handed pattern that was removed. This fact was established by the

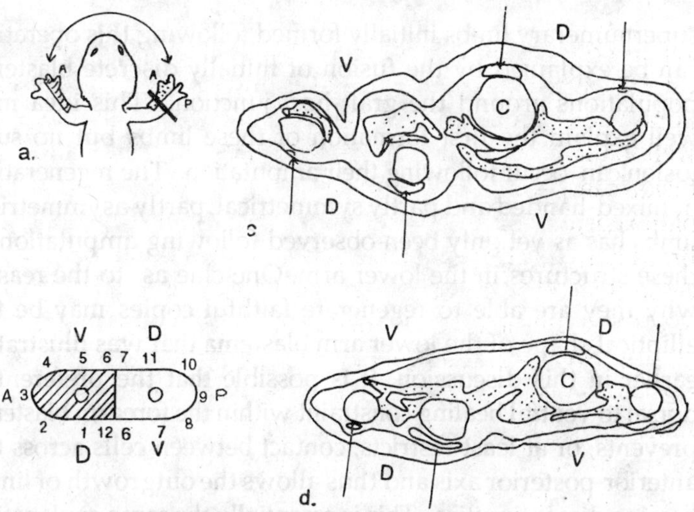

Fig. 11.9. (a) Diagrammatic representation of the operation performed to create mixed handed lower arms. The line across the limb represents the amputation plane. (b) Representation of the amputation plane of such a limb. Hatched area is the graft. (c) Camera lucida drawing of a histological section cut through the proximal metacarpal level of a surgically constructed mixed handed limb. (d) Camera lucida drawing of the regenerate formed following amputation of the limb shown in (c) after one month of healing. The level of section is comparable and the mixed handed pattern in the regenerate is closely similar to that initially formed following surgery. No extra structures were evident and a clear structural discontinuity still exists in the regenerated limb. The ventral muscles in (c) and (d) are marked by dots and the dorsal muscles are arrowed. C, cartilage; D, dorsal; V, ventral.

analysis of muscle patterns in the regenerates. Thus in these surgically created cases, as in the supernumerary limbs, clear structural discontinuities in the limb are maintained during outgrowth and apparently no attempt is made to smooth out the discontinuity. No extra pattern elements are found in these regenerates. The faithful regeneration of mixed-handed limbs also occurs following amputation of such limbs formed following ipsilateral blastema rotation. Maden has recently suggested that the complete range of anatomies in supernumerary limbs initially formed following this operation can be explained by the fusion of initially discrete blastema populations around the graft-host junction. This idea may well explain the first formation of these limbs but no such fusion can occur following their amputation. The regeneration of mixed-handed and partly symmetrical, partly asymmetrical limbs has as yet only been observed following amputation of these structures in the lower arm. One clue as to the reason why they are able to regenerate faithful copies may be the elliptical shape of the lower arm blastema that was illustrated earlier in this discussion. It is possible that the preferential dorsal to ventral healing constraint within the forearm blastema prevents, or at least restricts, contact between cells across the anterior-posterior axis and thus allows the outgrowth of limbs bearing discontinuities. This is essentially the same explanation as that put forward to explain the lack of a healing time effect following amputation of double anterior and double posterior axolotl forearms. It is important, therefore, to examine the regenerative potential of mixed-handed tissues constructed in the upper arm where the radial healing mode should ensure that the discontinuities in the pattern are smoothed out by intercalation. In these experimental limbs, extra structures would be predicted.

This particular category of results reveals that we are yet far from an adequate explanation for the cellular basis of pattern regulation in urodele limbs. It is clear, however, that the use of formal models that are clearly predictive remain essential as tools with which to further our knowledge. We must proceed with the hope that in the near future enough data will be available to allow the move to the analysis of pattern regulation at the cellular level. A clear understanding of this complex problem will only emerge when information from these different levels of organization can be held together. The urodele limb may be one of the systems in which this daunting task will be achieved.

12

Early Morphogenesis of Reptile Limbs

It is known that in reptiles certain somites in the anterior part of the trunk (namely, somites S_6 to S_{13}, inclusive, S_1 being the first post-otic somite) each send a long, ventral process into the forelimb rudiment. According to these authors, at least some of these processes (the five posterior ones) provide the forelimb rudiment with cells which will ultimately form the limb musculature. An analogous phenomenon occurs in the hindlimb rudiment. These observations have been supported by those of Milaire (1957) on chelonian embryos *(Emys, Pseudemys)* and lizard embryos *(Mabuia, Chamelaeo)*. During the last 15 years, limb morphogenesis in reptiles, and, in particular, the contribution of the somites, has been studied at our laboratory at Sannois, using embryos of the slow-worm *(Anguis fragilis)*, the green lizard *(L. viridis)* and chelonians *(Testudo graeca* and *Emys orbicularis)*. The first histological studies, carried out in 1962 on the forelimb rudiment of the slow-worm, suggested that the somitic processes initiate later morphogenetic changes in the limb rudiment. Studies carried out with batrachian embryos and experiments with avian embryos established the inductive role of the somite in the formation of the limb. This evidence encouraged us to undertake work on this question in the reptile embryo.

Effects of Somite Removal

In view of the difficulties involved in experiments with early stages due to the close apposition of the vascular area to the inner surface of the shell of the egg in many reptile species, experiments with embryos of the green lizard *(L. viridis)* were carried out on trunk segments explanted *in vitro* (a total of 30 explantations). At the 20-23 somite pair stage, some somites of the S_6 - S_{13} series were surgically removed, and during the following days it was ascertained that the somatopleure did not thicken alongside the site of the excised somites, and that no limb rudiments were formed. By contrast, 4 days after the operation, the somatopleure thickened alongside those somites of the series left in place, and from the sixth day of culture an outgrowth was produced at that point, in which an apical ridge differentiated, i.e., a limb rudiment was formed. The control for this experiment, consisting of the replacement of the excised somites, was carried out, but these somites did not form any ventral processes and the limb rudiment did not develop. Attempts at removal of the somites were carried out *in vitro* on embryos of *E. orbicularis*, but only partial removal was achieved, accompanied by damage to the somatopleure, so that no definite conclusions could be drawn.

Somitic Processes

In all the reptile species with well-developed limbs that have actually been studied, there are eight somites that send a process into the embryonic forelimb rudiment. By contrast, in reptile species with rudimentary or absent limbs the number of somites forming ventral processes is less than eight. Since the somatopleure thickens and grows out alongside each somite with a ventral process, the initial cranio-caudal length of the outgrowth of the limb rudiment is proportional to the number of somites forming such processes.

The reduction in the number of somites seen in embryos of the serpentiform reptiles is accompanied, in the species in which it has so far been possible to study this question, by various deficiencies : early degeneration of cells in the distal extremities of the somitic processes; weak somatopleure proliferation, and, furthermore, the path of the somitic processes may be abnormal and their cellular contribution weak (in the case of *Scelotes gronovii*.

TABLE 12.1

Number and Identity of Somites Sending Processes into the Forelimb Rudiment

Species	*Number of somites*	*Somites forming processes*[a]	*Authors*
Lacerta viridis	8	S_6 to S_{13}	Corning (1900) Raynaud & Vasse (1968)
Lacerta agilis	8	S_6 to S_{13}	Corning (1900)
Lacerta muralis	8	S_6 to S_{13}	Mollier (1895)
Lacerta sicula	8	S_6 to S_{13}	Raynaud & Vasse (1968) Vasse & Pieau (1970)
Testudo graeca	8	S_6 to S_{13}	Vasse & Pieau (1970)
Emys orbicularis	8	S_6 to S_{13}	Vasse (1973)
Scelotes gronovii	7-8	S_6 to S_{12} or S_6 to S_{13}	Raynaud & van den Elzen (1976)
Scelotes brevipes	5	S_6 to S_{10}	Raynaud, Gasc, Vasse, Renous & Pieau (1974)
Ophisaurus apodus	4-5	S_6 to S_9 or S_6 to S_{10}	Rahmani (1974)
Anguis fragilis	4-5	S_6 to S_9 or S_6 to S_{10}	Raynaud & Vasse (1968)
Tropidonotus tessellata	0		Raynaud (1972a)
Vipera aspis	0		Raynaud (1972a)

[a] S_1 being the first post-otic somite.

It has been proposed, as a working hypothesis, that the reduction in the number of somites sending processes into the somatopleure of the limb rudiment could be connected with the increase in the number of somites produced in the embryo of serpentiform reptiles. On the other hand, this reduction in the number of processes and the deficiencies that result could cause a functional deficiency in the mesoderm of the limb rudiment, a deficiency which in turn would entail the sudden or gradual involution of the apical ridge, or its weak differentiation. In fact, in all the embryos of serpentiform reptiles studied, it has been found that the apical ridge degenerates precociously. This was first noticed in the slow-worm embryo, where the apical ridge degenerates 8 days before the stage of embryonic development at which regression normally occurs in the embryos of lizards with well-developed limbs. This involution is due to the appearance within the cells of the ridge of centres of intracytoplasmic degeneration. This type of sudden and premature degeneration of the apical ridge has also been observed in embryos of *Ophisaurus apodus*. In other species, as is the case in embryos of *Scelotes inornatus*, *S. brevipes* and *S. gronovii*, regression of the apical ridge is slower or the ridge is only very slightly differentiated.

This interpretation, therefore, locates the initial cause of failure of limb outgrowth in the somite. This somitic deficiency probably induces a mesoderm deficiency, which in turn involves the regression of the apical ridge. This sequence of morphogenetic processes is supported by the observations reported above; it constitutes a working hypothesis and experiments have been undertaken to verify its accuracy. However, one could, *a priori*, envisage other possibilities, such as a genetic alteration in the cells of the apical ridge or the loss of capacity for response by the mesoderm cells, etc.

Following the involution of the apical ridge, there occurs in all species of serpentiform reptiles, a more or less complete cessation of development of the limb rudiment or even its total regression. Thus, in the slow-worm *(A. fragilis)*, 48h after the beginning of the marked degeneration of the apical ridge, the limb rudiment ceases to grow. Then, 2 days later, many cells degenerate in the mesoderm and the ectoderm, and the whole of the rudiment finally disappears. The same is true for the forelimb rudiments of the Glass snake and those of the three species of *Scelotes* studied. On the other hand, in the hind-limb rudiments, the regression of the ridge is slower or incomplete, and here a rudimentary limb forms which persists in the adult (as *in S. brevipes, S. gronovii:* These later observations add considerably to those made initially on *A. fragilis,* and support the hypothesis proposed earlier namely, that the spontaneous degeneration of the apical ridge of the limb rudiment constitutes, in the serpentiform reptiles, an essential factor in the arrest of limb development (in line with the ideas of Saunders (1948) and Cairns (1975) for the avian embryo).

Cellular Contribution of Somites to Limb Morphogenesis

The long somitic processes which penetrate the mesoderm of the limb rudiment in reptiles release somitic cells into the rudiment.

Electron microscope studies have made possible the tracing of the fate of the somitic processes and the way in which somitic cells are released into the limb rudiment. These ultrastructural studies have shown that in early embryos of the slow-worm *(A. fragilis)* and the green lizard *(L. viridis),* differences, which are constant at these stages, exist between the cells of the mesodermal somatopleural thickening and the cells of the somitic processes. The latter have thin mitochondria which an electron-dense matrix, and they contain numerous lipid droplets. The somato-pleural mesodermal cells have larger mitochondria which are less electron-dense than those of the somitic cells, and lipid droplets are rare.

On the basis of these criteria, we have found that :

(a) The processes of somites S_6 and S_7 push into the anterior part of the rudiment and terminate in a vesicle; this opens and the somitic cells detach from its wall and spread throughout the limb rudiment; this occurs in both the slow-worm and the green lizard;

(b) In the slow-worm embryo (when the allantoic bud is 0.7—1 mm in length) the prolongations of somites S_8 and S_9 are massive cellular processes which penetrate deeply into the limb rudiment and bend back towards the ectoderm ; but between them and the apical ridge is always a layer of two or three strands of mesodermal somatopleural cells. The distal extremity of the somitic process breaks up into blocks of cells, which in turn dissociate, thus liberating somitic cells which are distributed in several rows below or on the sides of the somatopleural cells positioned under the apical ridge. In the cellular masses resulting from the breaking up of the somitic processes and in the isolated somitic cells resulting from the dissociation of the blocks, the specific somitic characteristics of the cells are retained. The thin and dense mitochondria and abundant lipid droplets of the somitic cells enable them to be distinguished from the neighbouring mesodermal somatopleural cells. During the following days (allantoic bud length 1-2 mm), the somitic processes associate with the previous ones, now forming a large somitic mass along the axis of the projecting part of the limb rudiment. However, the specific somitic cell characteristics begin to disappear: they are less rich in lipids and the mitochondria become rather less dense. One can still, nevertheless, distinguish them from the somatopleural cells.

(c) In the green lizard embryo aged 1 to 3 days (incubation at 25°C), the somitic processes from somites S_6 and S_7 behave like those of the slow-worm. The processes of somites S_8 to S_{13} do not penetrate as deeply into the limb rudiment as to those of the slow-worm; but their distal part dissociates to the same extent, releasing somitic cells which mingle with the somatopleural cells, the cellular layers situated beneath the apical ridge being, as in the slow-worm, exclusively made-up of somatopleural cells.

A study of more advanced developmental stages is in progress. It should be added that the cells released by the somitic processes into the limb rudiments are similar in their cytological characteristics to the cells of the dermatome of the corresponding dermo-myotomes. Up to the 6 mm allantoic bud stage in the slow-worm embryo and up to the sixth day of incubation at 25° C in the green lizard embryo, differentiation of myofilaments has not been observed in the cells sent out by the somitic processes. Thus, there are not cells derived from the myotome. Perhaps, at these stages, they are still undifferentiated and will subsequently be able to develop into myoblasts. In green lizard embryos aged 6 to 7 days, the cells released by the somitic processes are distributed in the limb rudiment with a degree of alignment, forming strands which occupy the future position of the muscle blocks in the dorsal and ventral walls of the limb bud.

Cellular Constituents of the Limb Rudiment

This information has been obtained by means of histochemistry, cytophotometry and autoradiography.

Apical Ridge

Histochemistry and cytophotometric determinations show that at early stages of development *(L. viridis* embryo aged up to 2-3 days, incubation at 25° C; *A. fragilis* embryos at stages with the allantoic bud 0.7 — 1.5 mm in length), the nuclei of the apical ridge cells have a higher protein and nuclear RNA content than ectodermal cells situated on either side of the apical ridge. This relative increase rapidly falls and disappears after a few days; it indicates a specific metabolic activity in the cells of the apical ridge at an early stage in its formation.

Mesoderm and Somitic Processes

Histological and ultrastructural studies have shown that the mesoderm of the limb rudiment is made up of two lines of cells, one originating from the somatopleural mesoderm, the other from the somites. At the beginning of penetration of the somitic processes, it is possible to recognize these two cell lines, but at later stages the differences between them are less marked.

In *L. viridis and A. fragilis* embryos, the cells of the somitic process and the mesoderm cells are rich in proteins and RNA.

The specific configuration of the forelimb rudiment in the slow-worm, in which only the dorsal half eventually projects, allows one to follow the changes in the mesodermal cells situated in the neighbourhood of the somitic process and the apical ridge. The histological methods used were : Fast-green staining for protein; Unna-Brachet staining, and staining by the gallocyanin-Chrome alum method (±digestion of the RNA with ribonuclease) for the nucleic acids. These methods, together with cytophotometric measurements, showed that the somitic processes of somites S_8 and S_9 and the mass of mesodermal cells situated between the somitic processes and the apical ridge (i.e., in the dorsal projecting part of the limb rudiment) are particularly rich in proteins and RNA. The nuclei of these cells are larger and have large nucleoli with more RNA than the nuclei of the mesoderm cells in the ventral part of the rudiment. Moreover, there are extensive areas of RNA in the cytoplasm. Also the nuclei of the mesoderm cells of the dorsal half of the limb rudiment have a higher protein content than those of the ventral half (the relative increase can reach 28% at the 0.5-0.7 mm allantoic bud stage and 12-45% at 1 mm bud length). In addition, the nuclei of the mesoderm cells of the dorsal half of the rudiment have, at 0.7 - 1.5 mm bud stages, an RNA content two or three times higher than nuclei of the mesodermal cells of the ventral half. It is as if the mesodermal cells of the dorsal half of the limb rudiment have received a specific stimulation either from the somitic process or from the apical ridge, a stimulation which did not extend to other mesodermal cells of the ventral half.

On the other hand, in the extremities of the somitic processes (S_8-S_9 process in the slow-worm embryo and S_8-S_{12} processes in the green lizard embryo), cytophotometric determination show that the total protein content of the nuclei is a little less than that found in the overlying mesodermal cells. The difference is of the order of 10-20% in the slow-worm (1-3 mm allantoic bud) and 10-30% in the green lizard (aged 5-6 days, incubation at 25°C). At these stages, the nuclear protein content of the somitic processes is slightly higher in the slow-worm than in the green lizard. Finally, in these same stages, the nuclear protein content of the mesodermal cells (of the 4-5 cell layers situated beneath the apical ridge) is slightly higher in the slow-worm embryo than in the green lizard embryo (the values expressed in arbitrary units are 102 (1.5 mm allantoic bud) to 108.4 (1.5-1.8 mm bud) for the slow-worm, and 89 (at 5 days) to 95 (at 6 days) for the green lizard.

Autoradiography has given further information on the metabolic activity of the cellular constituents of the limb rudiment of the slow-worm and the green lizard. RNA and leucine-labelled protein synthesis is always high in the cells of the somitic processes. Evaluation of labelling intensity by counting silver grains in the nucleus in sections corresponding to those in the experiments cited above has shown that after treatment with [^{3}H]uridine the cells of the somitic processes are at first a little less heavily labelled than the mesoderm cells overlying them (0.8-1 mm allantoic bud stages). Then the labelling becomes equal in the somitic processes and the mesoderm (1-1.5 mm allantoic bud). After the degeneration of the apical ridge, the somitic process is slightly more heavily labelled than the mesoderm. This seems to be essentially due to a reduction of labelling in the non-somitic mesoderm cells. After the administration of [^{3}H]leucine to the embryo, the labelling was the same in the somitic process and the mesoderm (0.8-2.2 mm allantoic bud).

In the newly differentiated apical ridge in both the slow-worm embryo and the lizard embryo, there is little mitotic activity, and RNA synthesis is significantly reduced, whereas it continues actively in the cells of the ectoderm of the limb on either side of the ridge. The mesoderm cells show intense RNA synthesis. This synthetic activity continues to later stages in the green lizard, but, by contrast, it falls in the slow-worm embryo about 48h after the degeneration of the apical ridge. It is interesting to note that it is at this stage that the histological and ultrastructural signs of regression of the mesoderm cells of the dorsal half of the rudiment appear in the slow-worm (namely, reduction of nuclear and nucleolar volume, reduction of basophilia, swelling of the mitochondria). This suggests a diminution of activity of an apical ridge in which a large number of cells have degenerated. On the other hand, there was no sign of a reduction in RNA synthesis in the mesoderm cells of the slow-worm limb rudiment before the stage at which degeneration of the apical ridge occurs.

After labelling the embryo with [^{3}H]leucine, some protein synthesis is found to persist in the slow-worm apical ridge before it degenerates (whereas protein synthesis has become very much reduced at corresponding stages in the green lizard) and even 36h after the beginning of degeneration of the ridge, protein synthesis persists in the surviving cells.

With tortoise embryos *(T. graeca* and *E. orbicularis),* Vasse obtained results in studies on the apical ridge comparable to those obtained with the green lizard embryo. There is a very rapid reduction of DNA, RNA and protein synthesis in the apical ridge immediately after its formation. On the other hand, the fact that the labelling incidence for [^{3}H]thymidine in the mesoderm cells increases at the moment when the somitic process comes into contact with the somatopleure, suggests a stimulatory action of the somite on somatopleural cell proliferation.

13

Morphogenesis in Limbs of Calotes Versicolor

Work on the morphogenesis of the reptilian limb, or the other hand, is comparatively scanty and the literature scattered. Earlier workers focused attention on the ontogeny of autopodial skeletal elements, in order to reach phylogenetic conclusions. Casual analysis of reptilian limb morphogenesis is still in its infancy and, therefore, the descriptive data have to be discussed in the conceptual framework developed for chick limb morphogenesis.

There is considerable uncertainty as to the causative mechanism of limb initiation and the relative roles of the ectoderm, basement membrane, lateral plate mesoderm and somitic mesoderm in it. In *Calotes* the proliferation of the lateral plate mesoderm commences in the forelimb area at the 9-somite stage. At this stage the proliferating mesoderm is separated from the somitic mesoderm by intervening pronephric tissue, and from the basement membrane of the ectoderm by a considerable space. Later, in the 17-somite stage, the mesodermal cells fill the entire space between lateral plate and overlying ectoderm and come in contact with the ventral ends of somites. It is at the still later stage of 28 somites, that the ventral ends of somites penetrate the limb buds, and the ectoderm starts thickening over the limbs. Simultaneously the mesoderm cells become densely packed subectodermally and become loose proximally. A comparable situation is seen in the hindlimbs of *Calotes* and *Ophisaurus* and apparently in the forelimbs of the axolotl and *Emys*.

The observations favour the interpretation that the stimulus for the initiation of limb development is from lateral plate mesoderm and not from the ectoderm basement membrane or somitic mesoderm. The experimental evidence on this point needs to be carefully assessed since it is extremely difficult to separate cleanly the early embryonic tissues; also the potentiality of cells in early embryos continuously keeps altering, and there may be traumatic and other effects of experimentation on the tissues.

Wolffian Crest

The presence of Wolffian crest has been reported in certain reptiles and has been considered a general feature of amniotes. In *Calotes*, however, it could not be discussed at any stage of development.

Comparative Limb Growth

The hindlimbs in *Calotes* are not discerned in the 17-somite embryo but are seen in the 28-somite embryo. They initially develop faster than forelimbs, but both limbs are equally developed from stage 29 onwards.

Growth of Limb Mesoderm

The contribution of somitic mesoderm to limb mesoderm was at one time very controversial. The recent evidence not only in amphibians and reptiles, but also apparently in chick however, clearly indicates that somitic mesoderm contributes to proximal and central limb mesoderm, and further that this contribution is essential for normal development. Apparently the number of somites and also the extent of their penetration is greater in forelimbs than in hindlimbs. The somitic contribution to limb morphogenesis is reviewed by Raynaud (this volume) who concludes that the reduction in the development of limbs in reptiles is a consequence of deficiency of formation or function of the somitic processes.

The report that neural crest cells contribute to forelimb mesoderm in the mole remain unsubstantiated in any other species, including *Calotes*.

Table 13.1
Number of somites contributing mesoderm in reptilian limb development

Species	*No. of somites contributing to forelimb*	*Reference*	*No. of somites contributing to hindlimb*	*Reference*
Calotes versicolor	5(S7-S11)	Mathur (1974)	5 (S24-S28)	Mathur (1975)
Lacerta viridis	8(S6-S13)	Raynaud(1972); Raynaud & Adrian (1975)	—	
Anguis fragilis	4 (S6-S9)	Raynaud (1972)	3	Raynaud, cited in Rahmani (1974c)
Ophisaurus apodus	4 (S6-S9)	Rahmani (1974b)	3	Rahmani (1974c)
Scelotes brevipes	5 (S6-S10)	Raynaud et al. (1974b)	—	
Tropidonotus tessellata	0	Raynaud (1972)	—	
Testudo graeca	8	Raynaud et al. (1974b)	—	
Emys orbicularis	8(S6-S13)	Vasse (1974)	—	

The growth of limb mesoderm is asymmetrical and leads to a number of distinct limb asymmetries shortly after the initiation of limb development. For example in *Calotes,* in common with other vertebrates, the forelimbs begin to point postero-dorsally and hindlimbs postero-ventrally. The dorsal surface of both limb buds gradually becomes convex while the ventral surface remains rather flat (Zwilling, 1961). Both limbs show more rapid proximo-distal growth than cranio-caudal or dorso-ventral growth.

The distribution of limb mesoderm is *Calotes* at stage 30 is almost symmetrical around the long axis of the limb and comparable to that in chick at stage 18-21. Milaire, however, reports asymmetrical distribution of the mesoderm in reptilian and mammalian embryos. Observations in *Calotes,* as in chick, suggest that whereas the apical mesoderm contributes to the elongation of the limb bud, the compact subectodermal non-apical mesoderm contributes to the central mesoderm, causing it to be more compact.

APICAL ECTODERMAL RIDGE

Development of Apical Ectodermal Ridge

The thickening of general limb ectoderm in reptiles, as in other vertebrates, essentially begins only after the somatopleure mesoderm comes to lie subjacent to it, e.g., in the 28-somite stage in *Calotes.* The formation of apical ectodermal ridge (AER) begins at a still later stage (stage 29 + in *Calotes*). When fully formed (31 stage in *Calotes*) the AER, in external view of the embryo, appears as a ridge around the entire free margin of the limb bud and is nipple-shaped at the distal end of the bud.

In sections the maximally developed AER in *Calotes,* as in other reptiles and chick is fan-shaped; in some amphibians and mouse, the AER is apparently less developed. The arrangement of nuclei in layers gives the AER the appearance of being up to five cells thick in *Calotes.* Electron microscope studies on chick and man also indicate the multilayered nature of the basal cells (stratum germinativum) of limb ectoderm, even though some cells may touch both basal lamina and periderm. However, Ede, Bellairs & Bancroft (1974), after careful scanning and transmission electron microscopy on chick limb buds, conclude that the basal layer of the AER, as of the rest of the limb ectoderm, is only one cell thick and its stratified appearance is due to very close packing of cells so that the nuclei get lodged at various levels. This is an interesting point and its clarification in reptiles, as in other species, is awaited. However, Ede, Bellairs & Bancroft (1974), after careful scanning and transmission electron microscopy on chick limb buds, conclude that the basal layer of the AER, as of the rest of the limb ectoderm, is only one cell thick and its stratified appearance is due to very close packing of cells so that the nuclei get lodged at various levels. This is an interesting point nd its clarification in reptiles, as in other species, is awaited.

The mechanism of the formation of the AER is still somewhat obscure. Cell migration, cell division and changes in cell shape all play an important role. Changes in cell shape concern particularly the height of basal cells and microvilli in periderm cells, and are evident in reptiles as in amphibians and chick. The shape of both basal and peridermal cells apparently depends on their relationship with subjacent mesoderm and may have functional significance. Cell division in *Calotes* occurs not only in general limb ectoderm but also in the AER region (mitotic index 55 in forelimbs at stage 29-); Amprino (1965) considers AER cells to have a relatively poor mitotic rate in chick. Migration of ectoderm cells distalwards, relative to mesoderm, also contributes new cells to the AER region. The recent suggestion by Ede *et al.* (1974) that the cells of only the basal layer migrate whereas the periderm layer serves as a flexible cover is interesting.

The basement membrane invaginates to form a pit under the AER in *Calotes*, other reptiles and chick, but the pit is not reported in some amphibians and mouse. To assign an inductive

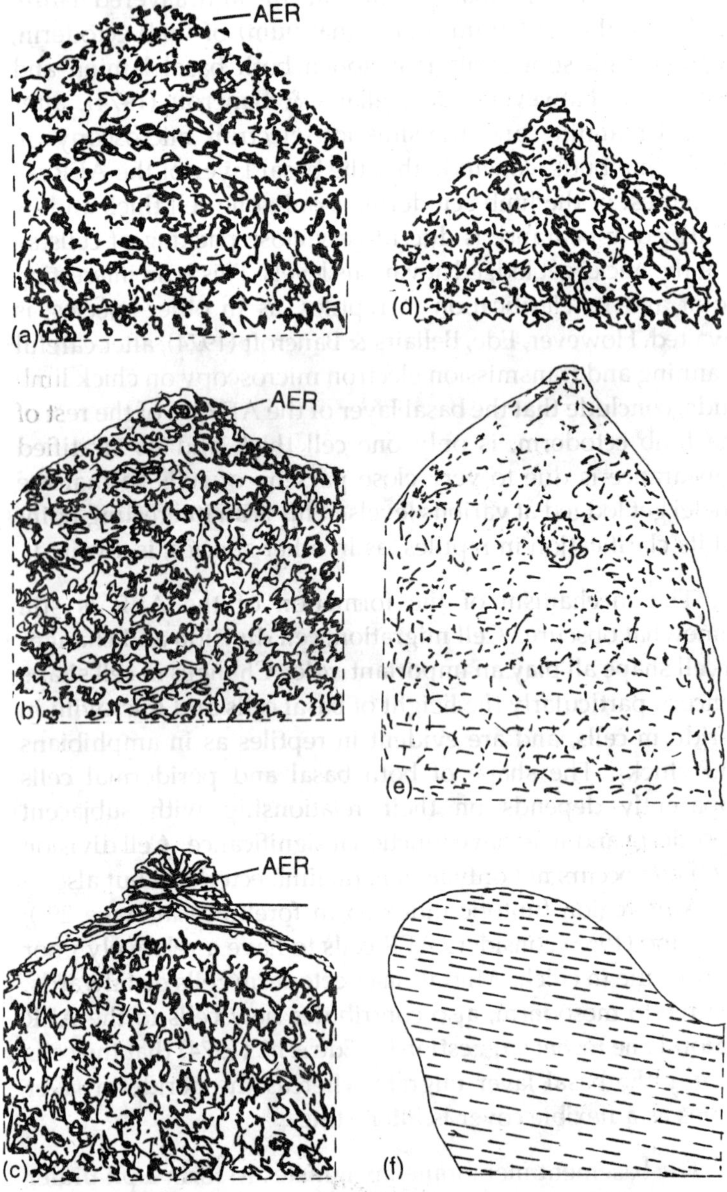
AER
(a)
AER
(b)
AER
(c)
(d)
(e)
(f)

Fig. 13.1. *Calotes versicolor* embryos. (a) Section of forelimb bud of a stage 29 + embryo to show the beginning of the apical ectodermal ridge (AER). The basement membrane is involuted under the AER. Note the two to four layers of nuclei in the AER region. Compared with other tissues in the limb, the AER is more pyroninophilic. (b) Section of forelimb bud of stage 31 embryo showing fully formed AER. The nipplelike appearance of the AER in section should not be confused with the nipple formed by the AER at the very distal end of the bud. The involuted basement membrane forms a pit under the AER. At this stage there is rich vascular supply to the limb. (c) Section of hindlimb bud of a stage 31 embryo to show fully formed AER. This is in all respects comparable to the AER of forelimbs. (d) Section of hindlimb bud of a stage 30 embryo to show the absence of alkaline phosphatase (calcium cobalt method) from AER and ectoderm. (e) Section of hindlimb bud of a stage 32 embryo showing the distribution of glycogen in AER. There is more glycogen in the involuted basement membrane and intercellular spaces, barely detectable amounts in cells of the AER and mesoderm. (f) Sections of hindlimb bud of a stage 32 embryo showing alcianophilia. The extracellular material of the condensation and the basement membrane (arrows) are more alcianophilic.

---

role to the pit or to the basement membrane is, therefore, premature. It may also be pointed out that electron microscope studies reveal that in *Lacerta* and the chick, but not in *Anguis*, the mouse or man, the basal lamina is discontinuous under the maximally developed AER.

Cytochemistry of Apical Ectodermal Ridge

The AER, compared with any other tissue of the developing limbs in *Calotes* (stage 29 + onwards), is more pyroninophilic, suggesting a markedly higher RNA content and, hence, more protein synthesis. Similar observations are already recorded for other reptilian and non-reptilian species. Moreover, Raynaud, Jeanny & Gontcharoff (1975) have observed higher levels of nuclear RNA and proteins in the AER, compared with other epiblast cells, in forelimbs of *Anguis fragilis* and *Lacerta viridis*.

Lacerta viridis

Alkaline phosphatase is missing from the AER and the remaining ectoderm of both fore- and hindlimbs of *Calotes* (Figure 2d; stage 29 + to 31 +). The distribution of the enzyme in the mesoderm is variable. A similar distribution pattern for the enzyme is described in the axolotl and *Xenopus* among amphibians, and in *L. viridis* among reptiles; in *A. fragilis*, the AER contains no more phosphatase activity than anywhere else in the limb. It has been claimed that in dogfish, chick and some mammals the enzyme is present exclusively in the AER; confinement to the AER is, however, contested for the chick. We, in disagreement with Milaire, do not assign any inductive role to the enzyme and are also unable to correlate the distribution of alkaline phosphatase with chondrogenesis in *Calotes* or other reptiles. It may be added that the alkaline phosphatases are a widely distributed, heterogeneous group of enzymes whose role in cell metabolism is not clear.

The distribution of glycogen (PAS-positive, α-amylase-digestible material) in limb buds apparently differs from species to species. In the limbs of *Calotes*, glycogen is seen in moderate amounts in the AER intercellular spaces (stages 31 and 32) and is barely detected in ectoderm or mesoderm cells. The presence of glycogen in the basement membrane and on the outer ectodermal surface (in all probability not in the peridermal cells) is not restricted to limbs but is found throughout the embryo; other authors have reported their observations on limbs only. Glycogen in various locations in limb ectoderm — basement membrane, basal layer, peridermal layer, AER — is reported in other reptiles, amphibians and mammals. We consider this is insufficient evidence to regard glycogen as an essential constituent of epidermal inductive system, particularly because of its presence in non-inductive locations like gut epithelium muscle and cartilage.

The presence of mucopolysaccharides (PAS-positive, α-amylase non-digestible) is also reported from the AER of *Calotes* two chelonian reptiles and the rat. Alciano philic acid mucopolysaccharides are also present in the basement membrane, peridermal surface, and particularly in cartilage anlage matrix, in the limbs of *Calotes* and chick.

The pattern of distribution of substances so far studied by cytochemical methods, apart from that of RNA, does not seem to provide any significant clues concerning the metabolism or morphogenesis of limbs in vertebrates. The presence of large quantities of RNA in the AER, on the other hand, may be regarded as reflecting the important role that the AER plays in limb morphogenesis.

Regression of Apical Ectodermal Ridge

With the formation of digital blastema, the AER regresses in the interdigital areas and becomes confined to the digital tips. In *Calotes*, and man this is evident from the height of the AER, while in the mouse it is reflected in the cytochemical properties of the AER in different regions. Later the formation of digits, the AER regresses over the entire limb.

The reason for the regression of the AER first in the interdigital and later on in the entire limb, is perhaps the lack of apical ectodermal ridge maintenance factor (AEMF) in the underlying marginal mesoderm. This is supported also by our observations that the AER, in trypan blue treated *Calotes* embryos, regresses whenever the necrosis is present in the distal mesoderm but not when it is in the deep mesoderm.

A different type of regression occurs in reptiles with rudimentary or no limbs (*Anguis fragilis, Ophisaurus apodus*). The whole AER and cells of somitic mesodermal origin degenerate, followed by the degeneration on non-somitic limb mesoderm. Electron microscopic studies revealed that the limb mesodermal cells of *Anguis* (whose embryos have rudimentary limbs), compared with those of *Lacerta viridis* (normal limbs), sendless numerous processes into the intercellular spaces and are less rich in mitochondria. Ede *et al.* (1974) observed similar differences between the normal and talpid wing bud mesoderm of chick.

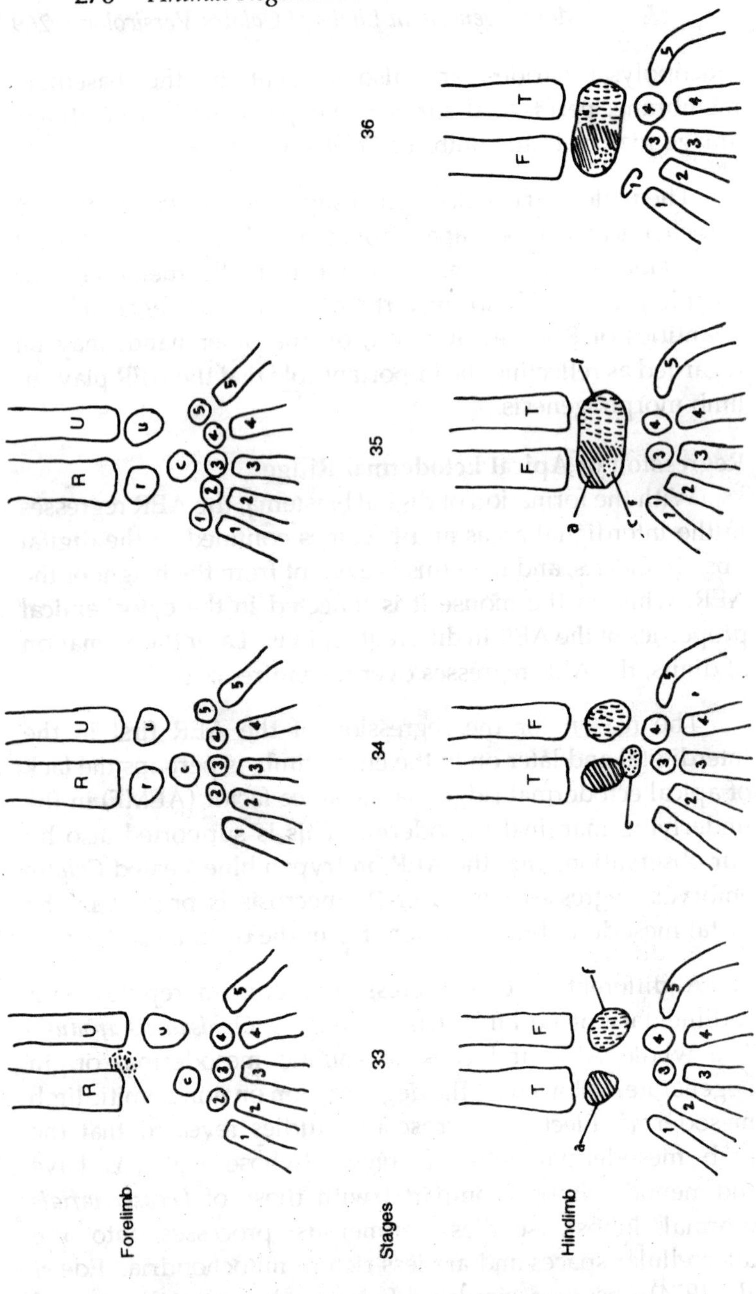
Forelimb
Stages
Hindlimb
33
34
35
36

Fig. 13.2. A diagrammatic representation of the ontogeny of carpus (forelimb) and tarsus (hindlimbs) in *Calotes versicolor*. In forelimbs the intermedium (i) appears as a mesenchymatous element at stage 33, and possibly undergoes necrosis later on. The other carpal elements are radiale (r), ulnare (u), centrale (c), and carpalia 1 to 5. In hindlimbs the centrale is seen as a distinct pre-cartilaginous element, though already attached to astragalus (a), only at stage 34. The fusion of astragalus, centrale and fibulare (f) to form the proximal tarsal element (T) appears complete at stage 35, but in the adult there are two calcification centres—one for fibulare, the other for astragalus/centrale — separated by a very thin uncalcified cartilage. Tarsalia 1, 3 and 4 are also seen.

Role of Apical Ectodermal Ridge

The relative roles of the ectoderm, particularly the AER, and the mesoderm in limb morphogenesis are a subject of controversy. Most workers in the field favour the Zwilling-Saunders hypothesis of limb morphogenesis. According to this hypothesis the AER is maintained by AEMF produced by post-axial mesoderm, and in turn induces outgrowth of limb parts in the responding mesoderm; the nature of the inductive action is still obscure. Available evidence in reptiles also favours the same hypothesis. For example: (a) whenever there is necrosis in distal mesoderm (in interdigital areas or trypan-blue-induced) the overlying ectoderm flattens, perhaps because of the absence of the AEMF; and (b) in the later stages the AER remains thick only on the digital tips, where mesoderm is still being added, thus indicating the role AER in outgrowth of the limb parts.

Chondrogenesis and Calcification

Patterns of chondrogenesis have been studied in *Calotes*, and some other reptiles. Our studies have clarified the issues involved with the homology and ontogeny of the carpal and tarsal elements in lizards. Interestingly these have also thrown some light on the question of relative roles of ectoderm and mesoderm in limb morphogenesis. For example, the pattern of chondrogenesis, such as the very late appearance of tarsalia

1, seems to be more consistent with the hypothesis that the AER, though important in limb morphogenesis, is probably not responsible for supplying information to the mesoderm concerning the number, size, shape and relative position of all the skeletal elements in *Calotes* limb.

TABLE 13.2

Homology of Tarsus in Lizards

Calotes versicolor	*Ascalabotes Fascicularis*	*Agama Colonorum*
(Mathur & Goel, 1976a)	(Sewertzoff, 1908)	(Holmgren, 1933)
Astragalus	Astragalus	'Centrale 1'
Centrale (distal)	Centrale	Centrale 2[a]?
—	—	Centrale 3[b]
—	—	Intermedium[b]
Fibulare	Fibulare	'Centrale 4'
Tarsalia	Meniscus	'Rudiment'?
—	Tarsalia 1[c] and 2[c]	Tarsalia 1[b] and 2[b]
Tarsalia 3 and 4	Tarsalia 3 and 4	Tarsalia 3 and 4

a According to Holmgren fuses with tarsalia 4.
b Fuses with astragalus (Holmgren's centrale 1>)
c Fuses with respective metatarsals. Homology is tentative elements marked [b] or are, in our view, non-cartilaginous blastema.

Astragalus, central and fibulare fuse to form a single proximal tarsal element.

Studies on epiphyses, sesamoids and calcification patterns in reptiles have already been reviewed and will not be discussed here.

Cell Death in Limb Morphogenesis

Recently cell death has been invested with an important role in the morphogenesis of organs. In reptiles the distribution of necrotic zones in developing limbs has so far been studied only in *Calotes*. Areas of cell death were studied by both histology and vital staining with neutral red; the results from the two techniques are generally comparable. Localised areas of cell death occur from stage 30 through 37. The anterior

Table 13.3
Homology of carpus in lizards

Calotes versicolor	*Ascalabotes fascicularis*	*Agama colonorum*	*Iguana tuberculata*
(Mather & Goel, 1976a)	(Sewertzoff, 1908)	(Holmgren, 1933)	(Holmgrem, 1933)
Radiale	Radiale	Radiale	Radiale
Centrale	Centrale	Centrale[2]	Centrale
-	-	Centrale 1[a]	-
Intermedium[b]	Intermedium[c]	Intermedium[c]	Intermedium
-	-	Centrale 3[c] and 4[c]	-
Ulnare	Ulnare	Ulnare	Ulnare
Carpalia 1	Carpalia 1	Carpalia 1[d]	(no data)
Carpalia 2 to 5	Carpalia 2 to 5	Carpalia 2 to 5	(no dat)

a Fuses with radiale.
b Undergoes necrosis.
c Fuses with ulnare.
d Fuses with metacarpal-1 (the adult does not have carpalia-1); possibly it is a non-cartilaginous blastema.

necrotic zone (stage 30 + through 33) and interdigital necrotic zone (stages 34 through 37) are extensive in both limbs. The posterior necrotic zone is ill-developed in forelimbs (stage 32) and is missing, or at best incipient, in the hindlimbs (stages 32, 33). There is no opaque patch. This pattern of necrotic zones is broadly comparable to that observed in the avian and mammalian limbs.

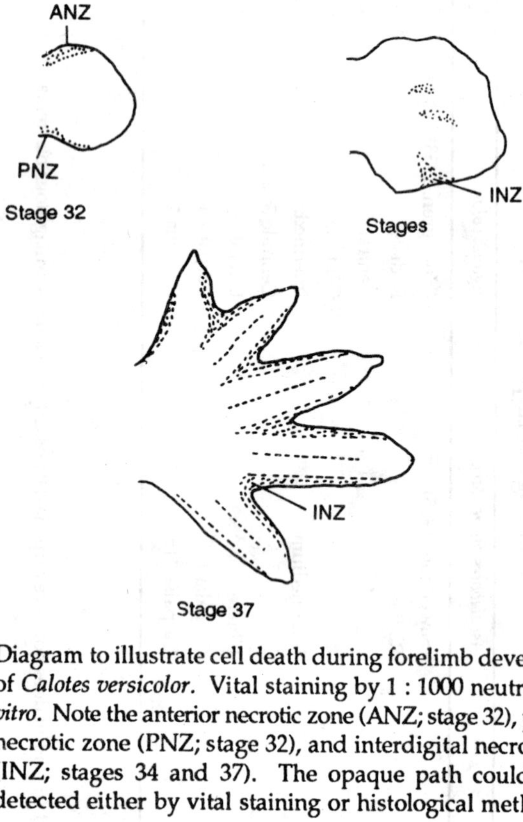

Fig. 13.3. Diagram to illustrate cell death during forelimb development of *Calotes versicolor*. Vital staining by 1 : 1000 neutral red in *vitro*. Note the anterior necrotic zone (ANZ; stage 32), posterior necrotic zone (PNZ; stage 32), and interdigital necrotic zone (INZ; stages 34 and 37). The opaque path could not be detected either by vital staining or histological methods.

Cell death in peridermal cells of AER has also been reported in reptiles, chick and mouse.

14

Growth and Determination in the Developing Limb

The characters of cells become determined in the course of growth; and the growth of cells depends on their state of determination. If we are to learn the mechanisms of determination, we must know the pattern of growth; and if we are to interpret the pattern of growth, we must know the states of determination. The first method to be described involves a forward projection of the pattern of growth step by step from the early bud. The second method, by contrast, gives a retrospective view; the parts of the differentiated structure are made to reveal directly the growth that they have undergone.

The first method was devised in order to work out fate maps for the early bud. It consists in calculating the trajectory of each part of the bud, using measurements of the local growth rate in each region at each successive stage; from this one can discover the cumulative amount of growth undergone by each part. The method is described in full elsewhere. The limb bud is idealised as an outgrowth elongating along its proximo-distal axis. Suppose that at time t we consider it to be composed of a large number of slices — for simplicity, let us say just three, of thickness l_1, l_2 and l_3 — and that the intrinsic growth rate in each slice has somehow been measured. We can then work out the amounts Δl_1, Δl_2 and Δl_3 by which the slices will increase their thickness in a time interval Δt, and plot the resultant displacements of the parts of the limb rudiment as it grows out. Repeating the procedure for successive time intervals, we can trace the successive positions of the parts, until we reach a stage at which their final destinies are clearly apparent. The prerequisite information as to the local intrinsic growth rate is culled in practice from measurements of mitotic index. Provided that the mitotic phase of the cell cycle, from prophase to telophase, has everywhere the same duration, the mitotic index in each region is proportional to the rate at which new cells are being born there. The main steps of the calculation are therefore done in terms of the number of cells in each slice, rather than in terms of its thickness. These numbers are easily related to slice thickness and position by some supplementary measurements and calculation.

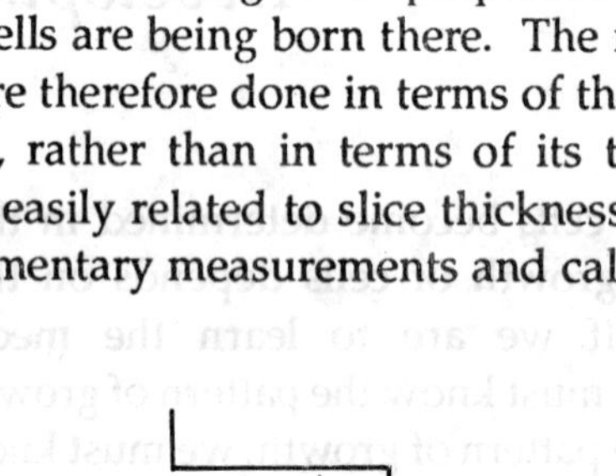

Fig. 14.1. Fate map calculation. The limb bud is subdivided, conceptually, into slices transverse to its axis of outgrowth. At time *t* the slices are, say, of thickness l_1, l_2 and l_3, and contain n_1, n_2 and n_3 cells respectively. Given the mitotic index in each of the slices, one can work out the numbers of additional cells. Δn_1, Δn_2 and Δn_3, that are born within each in the time interval t, and hence the thickness increments Δl_1, Δl_2 and Δl_3 and the resultant displacements. The procedure is repeated for subsequent time intervals.

The fate maps derived in this way have perhaps one or two special virtues in comparison with those obtained by other techniques, such as marking with carbon particles or [^{3}H]thymidine; they provide, for example, a more minute account of the early origins of the distal parts. But the interest of the calculation lies as much in a by-product as in the fate maps themselves. For in working out the shiftings of the cell population from the distribution of cell births, one discovers also how much each part of that population grows in numbers, and hence the number of cell generations corresponding to each period of development. That is, one establishes a natural time scale of development, a scale in terms of the cell divisions cycle, as opposed to an arbitrary time scale in terms of morphological stages, or the revolutions of the Earth. That time scale is of peculiar interest to us, because time plays a crucial part in the theory that we have put forward to explain how the limb gets its proximo-distal pattern of parts. Briefly; the parts of the limb are laid down in proximo-distal successions by proliferation of the cells in an undifferentiated region, which we called the progress zone, at the distal tip beneath the apical ectodermal ridge. We argued that the successive parts—humerus, forearm, wrist, etc.— owe their distinctive characters to a process of autonomous change in the state or 'positional value' of the progenitor cells in the progress zone. We speculated that the clock which sets the rate of the change might be the cell division cycle; for in that way the process of patterning the limb would always be harmoniously coupled to its rate of growth.

The fate map calculation supplies a titillating fact to encourage this line of thought. By cutting off the apical rise at different stages, one can discover when the parts of the limb are laid down, the whole process, as it turns out, from the emergence of humerus to the emergence of the terminal phalanges of the digits, lasts from about 18 to stage 28. The fate map calculation reveals that this interval corresponds to seven generations of cells in the progress zone(with an error of perhaps ± one generation). So how many generations of cells from the progress zone go to from one segment of the limb? There are five elongated segments of the limb, containing, respectively, humerus, ulna plus radius, metacarpals, proximal phalanges and distal phalanges. These are laid down in a fairly steady and regular progression, but in the middle of the process there is a striking anomaly: the rudiment of the carpal region takes about twice as long to emerge as the rudiment of a long segment, and has a correspondingly large initial size, inspite of the stunted and insignificant role for which it is destined. More will be said below about the peculiarities of the wrist; suffice it to say here, that we might reasonably regard it as the equivalent of about two long segments. The total number of limb segments then coincides, within the limits of experimental error, with the number of cell generations that is required in order to lay those segments down. One generation of cells from the progress zone constitutes one segment of the limb. If the intrinsic character of those cells near the tip of the limb is becoming progressively more distal with time, it must be doing so at a mean rate of one segment per cell cycle. Perhaps this is merely a numerical coincidence; perhaps it is a clue to a mechanism.

Differential Dilation

All this, however, deals only with the very grossest level of organisation of the limb. If we are to go further, and study not merely the succession of segments, but the development of the tissues within those segments, other methods are needed. As differentiation proceeds so does differential growth, bringing complicated internal distortions which cannot be analysed in terms of the simple one-dimensional scheme used in the fate map calculation. It becomes a painfully complex task even to describe the displacements in detail. There is, however, a simpler and more illuminating way to characterise the differential growth, and that is in terms of the volume dilations undergone by the parts. This information will not tell us directly where the parts of the developed structure came from;

but it will tell us *how big* their rudiments were. It is complementary to accounts of the elongation of the skeleton. We shall see that the limb contains not only some large things that spring from unexpectedly small beginnings, but also some small things that spring, or rather flop, from unexpectedly large beginnings.

The methods used here to measure the differential dilations is simple in outline. At the beginning of limb outgrowth, up to about the end of stage 22, the mesenchyme is of nearly uniform density and proliferates at roughly the same rate throughtout the bud. Thus, if radioactive thymidine is given early, the density of incorporation is roughly the same in all regions of the mesenchyme. If autoradiographs are prepared immediately, the tissue is seen to be almost uniformly labelled. Suppose, however, that the uptake of radioactive thymidine is halted, and the embryo is then left to develop and differentiate in peace for several days before the limb is fixed for autoradiography. In this period some parts will expand greatly and their content of radioactive thymidine will be spread out and rarefied in proportion; other parts will expand less and so keep their label concentrated. The density of label seen each over each tissue in the autoradiograph will be inversely proportional to the dilation that the tissue has undergone; and the total amount of label over any component will be proportional to the volume of early mesenchyme from which that component derives.

The main practical problem, apart from the usual headaches of autoradiography, was to find a way of halting the uptake of radioactive thymidine by the chick embryo: it is not degraded or excreted, as it would be by an adult animal, and the kinetics of thymidine incorporation are such that a chase of cold thymidine dropped into the egg is ineffectual. I resorted eventually to grafting. The donor embryo was supplied with five repeated doses of thymidine spread over a period of about 9 h, so that practically all the cells of the limb bud mesenchyme became heavily labelled. The right wing bud was then cut off, and grafted on to an unlabelled host, matched for stage, in place of the host's own right wing bud. This was done at or before the beginning of stage 23, and the trauma of the operation was then slight, as could be checked later by comparison with the normal left wing of the host. The left wing bud of the donor was fixed for autoradiography at the time of the operation as a control for uniformity of the initial labelling. The total dose of radioactivity per donor embryo was 5 μCi, or in a few cases 15 μCi. The wings were fixed for autoradiography after three or four more days of incubation, at stages 29, 30 or 32.

Finally, the density of labelling in the autoradiographs had to be measured, This was done photometrically by dividing up the area of the section into about 200 grid rectangles and scanning each in turn, under dark-field illumination that lit up the silver grains, with a photomultiplier. Measurements were thus made on each of a series of longitudinal sections through the wing, spaced 100 μm apart. The numerical result to be presented below, and summarised in Table 14.1, are averages over five complete wings analysed in this way, at stages 29, 30, 30, 30, and 32 respectively. The results will be discussed in terms of volumes of tissues and their dilation.

Some typical autoradiographs produced in this way and photographed with dark-field illumination, together with adjacent sections of the same limbs stained with toluidine blue to show the pattern of differentiation. To put it in a nutshell, the most prominent parts in the autoradiographs are the parts which are the most neglected by investigators of limb development — joints, wrist and perichondrium; and the parts to which investigators of limb development pay most attention- muscle and cartilage — are the parts that are the most insignificant in terms of their labelling.

TABLE 14.1

Composition of Differentiated Wings, Expressed in Percentages of Total Tissue Distal to the Mid-elbow (Means Over Five Specimens, ± Standard Error)

		Volume(%)	*Label (%)*	*Relative dilation*[a]
Segments	Forearm	52.8±1.4	57.0±1.8	0.93
	Wrist	17.0±1.5	26.1±1.2	0.65
	Hand	30.2±1.8	16.8±1.6	1.80
Cartilage	Ulna	3.1±0.4	2.4±0.2	1.29
	Radius	1.6±0.2	0.8±0.1	2.00
	digits	1.8±0.3	1.0±0.1	1.80
Perichondrium	Ulna	1.1±0.1	2.7±0.3	0.41
	Radius	0.8±0.1	1.1±0.1	0.73
	Digits	1.4±0.1	1.0±0.1	1.40
Muscle		7.6±0.5	5.9±0.6	1.29
Elbow tissue with ≥ 3x mean label density		3.5±0.4	18.5±2.0	0.19

Relative dilation = (%)/label (%).

The heaviest labelling of all is to be seen at the elbow joint at the meeting place of the humerus, the ulna and the radius. The label at this centre is 10 times denser than the average for the limb. It might be thought that the nugget of heavily labelled tissue at the elbow is nevertheless so small as to represent only a very minor fraction of the original undifferentiated limb bud. Let us therefore compare it with the entire forearm, measured from the middle of the elbow to the articulation of the ulna and radius with the carpals. To be specific, let us delineate the heavily labelled nugget at the elbow as that region within which the density of label is at least three times greater than the average. The volume picked out in this way in the differentiated wings is indeed quite small: just 7% of the volume of the forearm. But the total

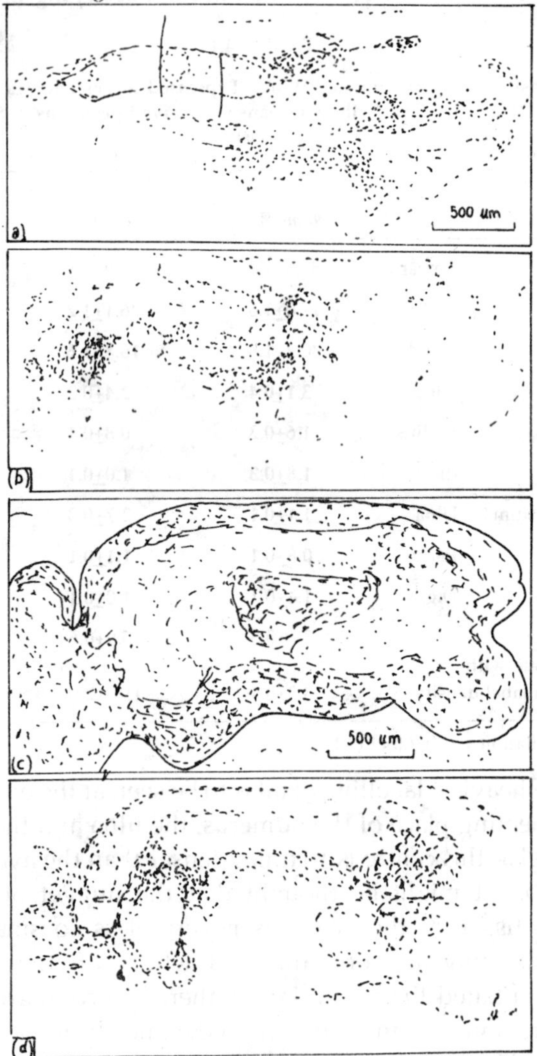

Fig. 14.2. (a), (b) Longitudinal section of wing fixed at stage 29. (a) Stained with toluidine blue; (b) autoradiograph showing differential dilation. (c), (d) Longitudinal section of wing at stage 30, passing through head of humerus, full length of ulna, distal end of radius, and wrist. The wrist was slightly flexed, so that most of the hand lies out of the plane of section. (c) Stained with toluidine blue; (d) autoradiograph. Note proximal boundary between labelled graft and unlabelled host.

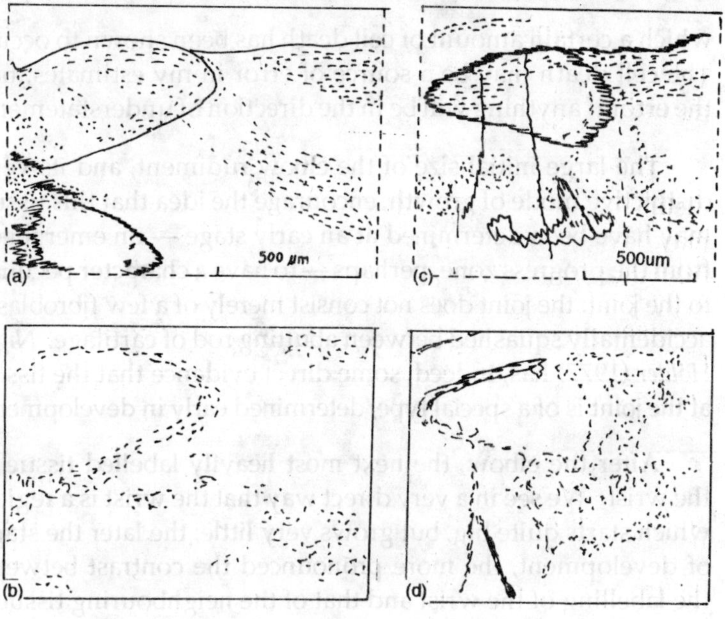

Fig. 14.3. ***(a), (b)*** **The neighbourhood of the elbow at stage 32, showing proximal ends of ulna and radius and some muscles of the forearm.** ***(a)*** **Stained with toluidine blue; (b) autoradiograph.** ***(c), (d)*** **Longitudinal section through wrist at stage 32.** ***(c)*** **Stained with toluidine blue; (d) autoradiograph. (The heavy transverse lines are creases in the sections.)** ***(e), (f)*** **Transverse section through humerus at stage 30.** ***(e)*** **Stained with toluidine blue; (*f*) (autoradiograph).**

quantity of label that it contains amounts to 32% of the total label in the forearm. Thus in terms of its origins it is the equivalent of one-third of the entire forearm: that should be its relative size on the early fate map. The heavily labelled tissue at the elbow probably corresponds to the 'opaque patch' in which a certain amount of cell death has been shown to occur. The cell death may be a source of error in my estimates; but the error if anything will be in the direction of understatement.

The large initial size of the elbow rudiment, and its very distinctive mode of growth, encourage the idea that this tissue may have been determined at an early stage — on emergence from the progress zone, perhaps — to have a character peculiar to the joint: the joint does not consist merely of a few fibroblasts accidentally squashed between abutting rod of cartilage. *Nigel Holder* (1977) has, indeed, some direct evidence that the tissue at the joint is of a special type, determined early in development.

After the elbow, the next most heavily labelled tissue is the wrist. We see in a very direct way that the wrist is a region which starts quite big, but grows very little; the later the stage of development, the more pronounced the contrast between the labelling of the wrist and that of the neighbouring tissues. We can compare the wrist with the hand, defining the wrist as all that lies between the distal ends of the ulna and radius and the proximal ends of the metacarpals, and defining the hand as everything distal to the wrist. At stage 30, the volume of the wrist is thus only just over half that of the hand; but the amount of label in the wrist is one and a half times greater than the amount in the hand. In the undifferentiated bud, therefore, the precursor of the wrist was one and a half times larger than the precursor of the hand.

We can use the autoradiographs to compare the sizes of the limb segments when they are first laid down, i.e., when they emerge from the progress zone; but we must allow for the fact that the distal parts emerge later than the proximal, and consequently emerge more dilutely labelled. According to the argument given above, the forearm represents one generation of cells, the wrist roughly the next two generations, and the hand all subsequent generations. If the volume of tissue emerging from the progress zone were the same in each generation, we should therefore expect to find in the autoradiographs of the differentiated limbs that the total amounts of label contained in forearm, wrist and hand were in the ratio 1:1/2 + 1/4:1/4, that is, 1:0.75:0.25. My measurements give instead the values 1:0.46:0.29. This implies that the volumes of tissue emerging from the progress zone in successive generations are not quite the same, but somewhat reduced during the generations of the wrist. This might be explained in terms of the changes that are then occurring in the dorso-ventral depth, antero-posterior extent and cell packing density in the apical part of the bud.

The peculiar pattern of growth at the wrist and elbow provides the solution of another small puzzle. It is a common observation that carbon parties inserted as markers in the early bud are very often found later to be concentrated in these two regions; and grafts of tissue to distal regions of early buds generally turn out to articulate with the wrist of the developed host limb. There are no mysterious attractive forces at work here; merely the geometrical consequences of the pattern of growth. The prospective elbow and wrist are disproportionately large parts of the early bud, and an object inserted at random into the early bud has a high probability of falling within one or the other. The joints of the digits are also sites of relatively heavy labelling; but the appearance is less dramatic in my autoradiographs, since the digits are less advanced in their differentiation, and the labelling of the hand as a whole is rather faint. Lastly, one can analyse the limb into the familiar types of differentiated tissue, and work out from the autoradiographs how much of the early bud has gone to form each of them. In the sections stained with toluidine blue, the muscle, the cartilage and the perichondrium, for example, can be quite easily identified. Of these, the perichondrium is the most conspicuous in the autoradiographs, but the most neglected in studies of limb development. The cartilage, by contrast, is very lightly labelled; and the muscle in the autoradiographs is barely distinguishable from the loose connective tissue that surrounds it.

The volume occupied by all these constituents of the stage 30 wing is relatively small; there is less cartilage and muscle than one's first visual impression would suggest, but more perichondrium. As for the amounts of label, the cartilage of the ulna plus that of the radius contributes a mere 6% of the label in the forearm as a whole — only one-fifth as much as the heavily labelled nugget at the elbow joint. The perichondrium of the ulna and radius, thin as it is, contains slightly more label than the cartilage — about 7% of the total in the forearm. The muscles in the forearm and distal parts of the wing contain only another 6% of the total label. These estimates are fairly precise and constant from one specimen to another, despite the differences of stage; this is evidence against late recruitment of surrounding undifferentiated cells into the muscle, cartilage and perichondrium.

From these figures it is clear first of all that it is misleading to discuss the differentiation of the wing bud chiefly in terms of a dichotomy between muscle and cartilage for .88% of the bud will differentiate in neither the one way nor the other. We should not be surprised that Dienstman *et al.* (1974) could get only about 10% of the cells in their cultures of limb bud mesenchyme to differentiate as muscle or cartilage; nor should be suppose, as they do, that the remaining 9% of their cells, that had not differentiated in either of these ways, are mostly to be classified as precursors of muscle or cartilage. It is almost certainly by differential growth that the muscle in the limb achieve its final prominence; its early rudiment is small. The labelling measurements show that the prospective perichondrium represents roughly the same volume of the early bud as does the prospective cartilage. By stage 30 or so, however, the cartilage has become relatively inflated, and the perichondrium appears to be merely a think and rather insignificant layer of cells squashed against its surface. But that is not the end of the story : for the perichondrium eventually takes its revenge, and produces the bone that replaces the cartilage. It is thus tempting to speculate that bone, although it differentiates late, might be determined early, on a par with cartilage.

A final lesson to be drawn from these measurements concerns the cartilage. In almost every study of pattern formation in the limb bud, this is the tissue which serves to mark the final outcome. Indeed, from reading some accounts one might suppose that the limb bud gives rise to nothing else. But that could scarcely be further from the truth. Of the undifferentiated forearm rudiment, only 4.2% goes to form the ulna, and only 1.4% to form the radius. Not only are the cartilage rudiments a tiny fraction of the whole, they also grow in an unrepresentative way. The radius rudiment, for example, by stage 30 has dilated by twice the average factor, that is, 20 times as much as the most heavily labelled tissue at the elbow. Bearing these facts in mind, we should not be surprised at the contortion and variability of the patterns of cartilage seen in limbs whose buds have been rudely cut apart and reconstituted from fragments pinned together.

Control of Growth

Finally, one may ask why it is that the different parts of the limb grow at different rates. One conceivable answer might be that their surroundings constrain them mechanically in different ways. But this will scarcely explain why the wrist, for example, grows so much less than the forearm, or the rudiment of the fibula so much less than that of the tibia. Much the most plausible explanation is that the cells in different places are in different states of determination — or in Wolpert's (1971) terminology, have different positional values — and that the states of determination define the rates of growth. The mechanism for assigning those different states of determination has long been a central concern in the study of limb development, and various schemes have been proposed. This is not the place to argue their merits, but rather to emphasise a point which is implicit in them all. They all suppose, and with good evidence, that the cells that are specified to form the different regions of the limb are somehow intrinsically different, even though they may give rise to the same histological types. The cells that form the cartilage of the humerus, for example, do not have quite the same character as those that form the cartilage of the ulna or of the carpals. The cartilage cells are alike in the eyes of the histologist, but they are somehow non-equivalent there is something different in the mode of control of their differentiation. The non-equivalence is of fundamental importance for the patterning of the limb, and it forces a revision of the old histological concept of cell type. Some of the plainest manifestations of non-equivalence may be seen in the differences of growth and maturation among the bones of the limb.

15

A Model for the Control of Limb Growth and Development

Experimental embryology has revealed the morphology and kinetics of limb growth and differentiation and their responses to microsurgical intervention. From these has developed a series of theories to account for the observed interactions; for example, Amprino's biomechanical control, Zwilling's maintenance factor, Saunders' outgrowth inductor and others. Recent work has led to a partial re-evaluation of these experiments and a new theory based on a permissive rather than a directive control. In all these theories, however, the precise identity of the controlling agents and their mode of action remain unresolved, and the interactions seldom lend themselves to simple molecular calculations.

Molecular biology, on the other hand, gives us a wealth of specific interactions, feedbacks and control mechanisms with which to direct growth, development and differentiation, but no way to apply them to achieve the temporal and spatial localisation of invisible biochemical switching that is cell determination. It is the purpose of this paper to outline a simple, biochemically realistic model, based on cell-to-cell diffusion and allosteric switching, that gives both the spatio-temporal localisation and the global interactions necessary to control growth and development in a multitude of systems. With sufficient refinement, the model will allow direct calculations of the molecular constants of the controlling agent or agents.

Basic Model

The basic model is given by Wilby & Ede (1975). This is the wave form gradient model, developed for the antero-posterior (AP) pattern of cartilage in the chick limb bud. There are four main steps in this model:

(i) a localised initiation of morphogen (M) synthesis, giving,
(ii) an increase in intracellular (M), modified by diffusion to neighbouring cells and, where conditions permit, leading to
(iii) a threshold concentration *T*, at which point destruction of M is initiated and, simultaneously, cell determination occurs. Subsequently
(iv) a decrease in intracellular (M) occurs, generating a 'sink, and eventually the cell differentiates.

The transitions between 'quiescent' (i.e., non-synthesising) cell, 'synthesising' cell and 'destructive' cell are irreversible, most probably involving changes in enzyme structure.

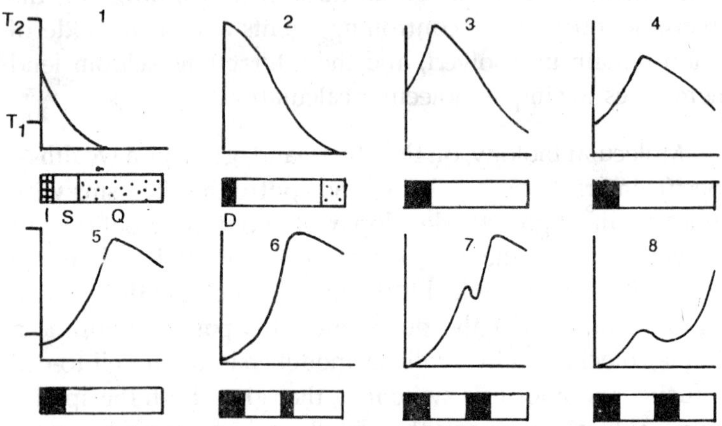

Fig. 15.1. Development of spatial localisation model. T_1, synthesis initiation threshold; T_2, destruction threshold; I initiator region; S, synthesising region; Q, quiescent region; D, destructive region. Heavy line is morphogen concentration. Note oscillation of peak between steps 3 and 6 and peak splitting in steps 6 to 8.

One-dimensional Model

In such a system in a one-dimensional cell string of fixed length, initiation' is achieved by setting a lower threshold T_1 for [M], below which the cell is quiescent and above which the switch to synthesis occurs. By putting a cell already synthesising M (the initiator) at one end of the string, free diffusion between cells ensures that the initiation event propagates at a constant velocity along the cell string, followed by periodic initiation of destruction at threshold T_2.

Two-dimensional Model

In two dimensions, this model can produce a reasonable simulation of the cartilage pattern in the chick limb bud if patterning is confined to a series of non-interacting thin sections. The practical implications of this are that there must be little or no diffusion of M posteriorly from the distal tip. An antero-posterior polarity develops from the local gradient dynamics of this model, rather than from any global control involving the zone of polarising activity (ZPA).

GLOBAL CONTROL MODEL

Theory

To develop a realistic global control model for limb development, the morphogen gradients must be used more efficiently; cell division rate is made a function of gradient concentration, cell orientation after division a function of gradient slope and cell death a threshold function, determined both by gradient concentration and differentiated cell state. Cell determination remains linked to the switch to morphogen destruction and cell-to-cell diffusion becomes a function of cell state, decreasing a hundredfold upon cell determination. In this way the requirements of the AP patterning model are met and the periodicity of patterning in the proximo-distal (PD) axis is eliminated, allowing for a more uniform control.

The initiation event must also be modified. Primarily it is contact-dependent, synthesis of M in limb mesenchyme being initiated by direct contact with cells of the apical ectodermal ridge (AER); thereafter, however, the initiated state is inherited on division. Other cell-to-cell interactions are as in the AP patterning model but, in addition, there is diffusion to the AER which acts as an external homeostatic sink, maintaining a constant low distal level of morphogen T_2 and allowing a stable gradient pattern to develop.

The One-dimensional Model

In a growing one-dimensional cell string. Initially the string is five cells long — three mesenchyme cells and two distal AER cells. On the basis of the scanning electron microscope studies of Ede, Bellairs & Bancroft (1974), the three mesenchyme cells are all considered to be in direct contact with the AER by filopods, and so are initiated to synthesis M. This three cell deep contact has important implications when considering the effects of AER removal and grafting; it may protect cells from dying in the presence of sub-death threshold levels of (M) and may re-initiate determined (but not differentiated) cells.

Cell Division

Cells to divide are picked at random at a level proportional to morphogen concentration: 1% (background) when [M] = 0.0, 10% when [M] = 10.0 (maximum concentration *T*). The effect being simulated is not a uniform variation in cell cycle length but a block at some point in the cycle (possibly G_1) at which cells may 'drop out' and differentiate.

Cell Orientation

With no preferred cell orientation, only a random 50% of dividing cells can be expected to contribute to string elongation. When a cell is so picked, an identical copy is inserted into the string on its distal side, displacing all distal cells one cell diameter. If there is to be cell orientation, this is achieved by increasing the probability that a dividing cell will contribute to orientation in proportion to the local morphogen gradient slope, providing this exceeds a pre-set minimum.

Cell Death

If, in an undetermined cell, the morphogen concentration falls below a death threshold T_d, that cell is assumed to die and is removed from the string. The gap is filled by displacing all distal cells one space proximally.

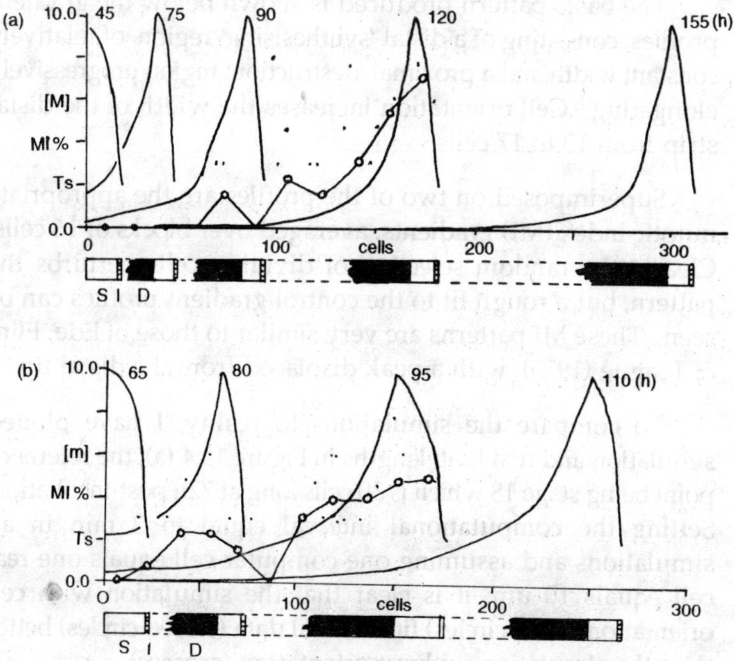

Fig. 15.2. Development of global control model for normal chick. *(a)* No cell orientation; *(b)* with cell orientation. [M], concentration of morphogen at indicated times (heavy lines); M1% mitotic index (small dots represent four cell average, linked open circles represent 16 cell averages); T_s, sink threshold ; S, synthesising region: D, destructive region; 1, AER sink/initiator, Critical slope for *(b)* is 0.2 units/cell diameter.

Normal Chick Simulation

The development of a cell string with no preferred orientation after division, up to about 300 cells long (the simulation limit is 360). The time base (top horizontal axis) of this simulation is converted from the computational interval, using the most realistic simulation to date as the reference. Figure 15.2 (b) shows the effect of cell orientation with the minimum slope set at 0.2. More cells in the distal regions then contribute to elongation, but more cells are also dividing in these regions, giving a faster growth rate.

The basic pattern produced is shown below the gradient profiles, consisting of a distal 'synthesising' region of relatively constant width and a proximal 'destruction' region progressively elongating. Cell orientation increases the width of the distal strip from 13 to 17 cells.

Superimposed on two of the profiles are the appropriate mitotic index (MI) gradients, averaged over blocks of 16 cells. Clearly the random selection of dividing cells perturbs the pattern, but a rough fit to the control gradient profiles can be seen. These MI patterns are very similar to those of Ede, Flint & Teague (1975), with a peak displaced from the distal tip.

To compare the simulations to reality, I have plotted simulation and real limb lengths in Figure 15.4 (a), the reference point being stage 18 which is 30 cells long at 72h post-incubation. Setting the computational interval equal to 3 min in all simulations and assuming one computer cell equals one real cell equals 10 um, it is clear that the simulation with cell orientation (open circles) fits the real data (closed circles) better than the simulation without orientation (crosses).

Talpid^3 Mutant Simulation

In the *talpid* (*ta*^{3}) mutant, limb morphology is drastically altered; the limbs are shorter, wider and polydactylous individual cartilage elements are shorter, especially proximally, and the mitotic index pattern is altered giving an initial proximo-distal gradient then the suggestion of a medical peak before developing a disto-proximal gradient. A number of observations based on cell and tissue culture indicate that one difference between ta^3 and normal may be in the degree of cell to cell connection and thus in the degree of cell-to-cell diffusion in a model such as mine.

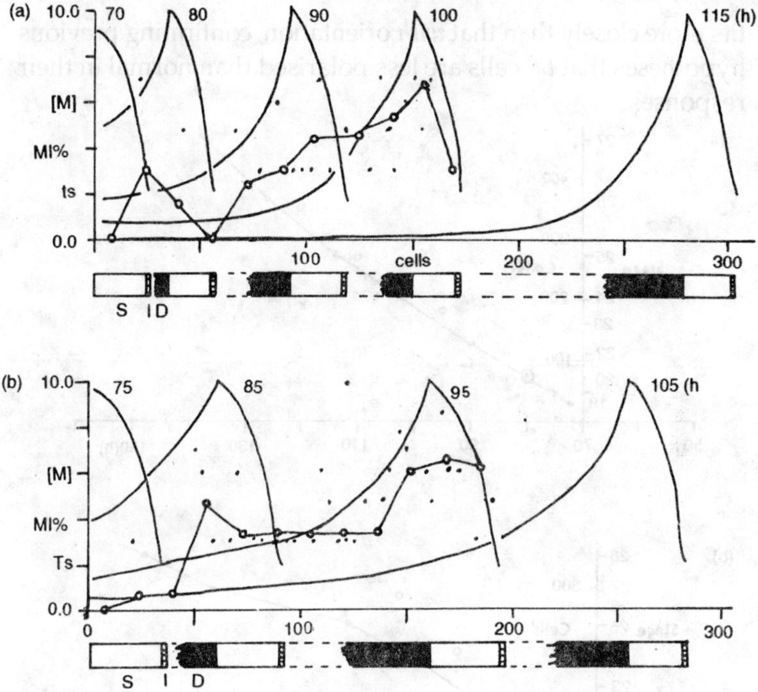

Fig. 15.3. Development of global control model for ta^3 mutant. All symbols as for Figure 15.2. Note change in gradient profile, size of synthesising area, mitotic index gradient and elongation rate, all resulting from a three-fold increase in cell-to-cell diffusion.

To simulate *ta*^{3} development I have therefore simply increased the diffusion rate constant three-fold from 0.1 to 0.3, and get the patterns. shown in Figure 15.3. Figure 15.3 *(a)* has no cell orientation; Figure 15.3 *(b)* has the same degree of orientation as Figure 15.2 (b), increasing the distal strip from 22 to 30 cells.

Two obvious differences can be seen between these and their 'normal' equivalent: the 'synthesising' regions are larger and the peak of cell division is displaced further from the distal tip. By comparison with real *ta*^{3} limb elongation data. (b) it can be seen that now the simulation *without* orientation fits more closely than that *with* orientation, confirming previous hypotheses that ta^3 cells are less polarised than normal in their responses.

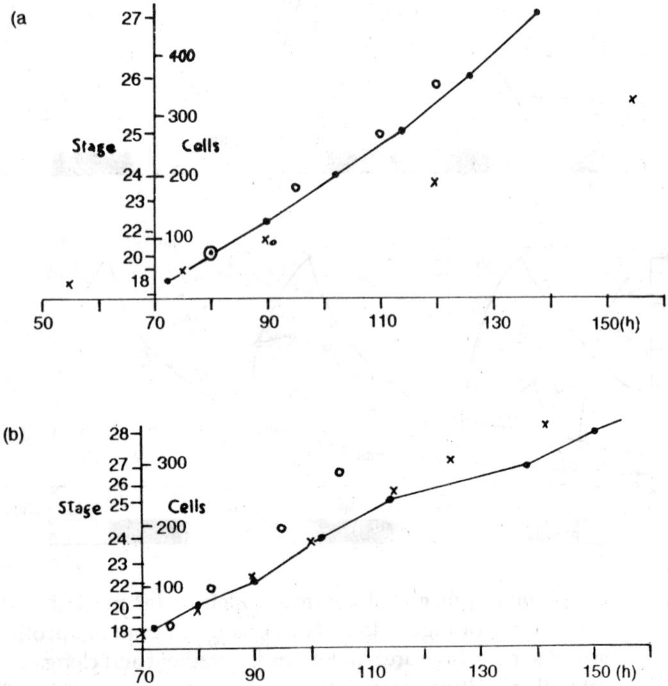

Fig. 15.4. Comparison of simulation data with real elongation rates. *(a)* Normal chick; *(b)* ta^3 mutant chick. Open circles, simulation with orientation; crosses, simulation without orientation: closed circles, real length. Measurements are based on the assumption that one computer cell = one real cell = 10 μm.

A one-dimensional simulation clearly cannot provide much information about the two-and three-dimensional development of the limb but, since the best ta^3 simulation has no cell orientation, the obvious conclusion would be equal PD and AP growth rates for a given region. This could clearly result in the typical fan-shaped ta^3 limb.

The Proximo-Distal Patterning Model

The utility of the growth control model is not confined to

simple elongation simulations, because it contains all the necessary information for a proximo-distal sequential pattern and can thus be used to simulate most of the microsurgical experiments on chick limbs to date. The PD pattern derives from the 'positional information' model of Summerbell and co-workers (Summerbell *et al.*, 1973; Summerbell & Lewis, 1975), the essence of which is that, within a narrow distal strip of mesoderm termed the 'progress zone' (PZ), the existence of which depends on the AER, an autonomous 'clock' linked to cell division specifies the determined state of the cells, which is fixed on exit from the PZ. The AER thus permits rather than directs the pattern. In its absence the PZ decays and determination ceases.

By equating the region of synthesis and high cell-to-cell diffusion in my model with the progress zone, I produce Figures 15.5 and 15.6 for normal and ta^3 respectively. These figures show several things:

(i) There is a progressive emergence of clones 1 to 5 from the PZ, these clones deriving from individual cells in stage 18 simulations.
(ii) Simulations can be precisely staged by developmental time so as to compare lengths under different conditions.
(iii) The pattern of number of divisions in the PZ (N_m) forms a ragged, but distinct, PD-increasing gradient.

The N_m Profile

Of vital importance to the PD patterning is the development of a smoothly increasing N_m value. In my simulations I have approximated this by averaging over blocks of 10 to 20 cells then drawing a straight line through the average points. In any system, since there is no mitotic synchrony, there *must* be some form of AP or PD averaging, the nature of which is still obscure.

Partially by comparison with published values (cell doubling times, in a uniform growth model, Summerbell &

Lewis (1975), based on AER removal experiments) and partially from a direct comparison of determined states from the above simulations and a range of more complex graft interaction simulations, I have constructed a comparison table for time, stage, length (real and simulation) regional determination, t and N_m. From this the patterns in Figures 15.5 and 15.6 are calculated. The most obvious difference between t and N_m is in the carpal region, reflecting both the uniform elongation rate that I find over stages 18 to 25 and my concept that one division interval equals exactly half a long bone (humerus, radius, ulna or metacarpal) or one short bone (carpal phalange).

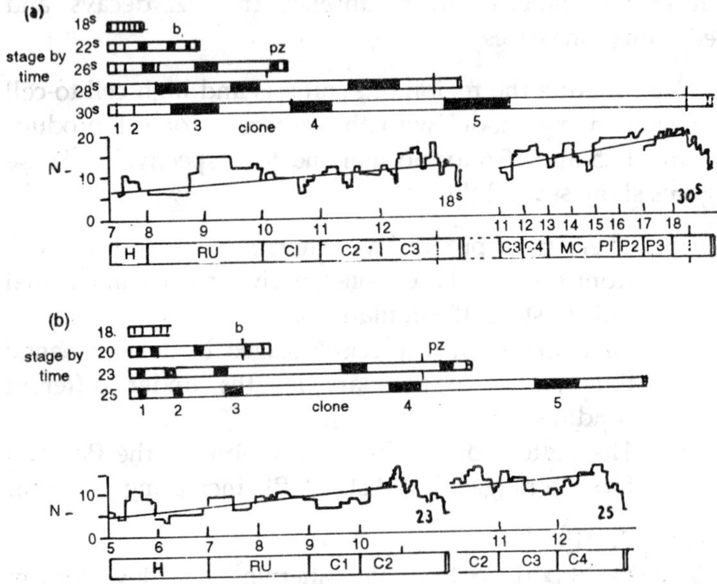

Fig. 15.5. 'Progress zone' model for proximo-distal patterning in normal chick. (a) simulation without cell orientation : (b) simulation with cell orientation. b, boundary of progress zone; pz, progress zones 1-5, progeny of single cells in stage 18. Superscript 'S' indicates that the simulation is *too short* for that developmental stage (defined by time). N_m number of mitoses that occurred within the PZ. Horizontal axis gives N_m intervals of linear average. H, humerus; RU, radius and ulna; C_1-C_4, carpals: MC, metacarpals; P_1-P_4 phalanges. Determination, differentiation pattern.

The origin of the τN_m differences lies in my use of a 'medial peak' gradient to control cell division rather than assuming a series of blocks of tissue each with a uniform division rate but graded disto-proximally. The result is that my version of the chick limb develops from fewer cells dividing faster than in the Summer-bell & Lewis model.

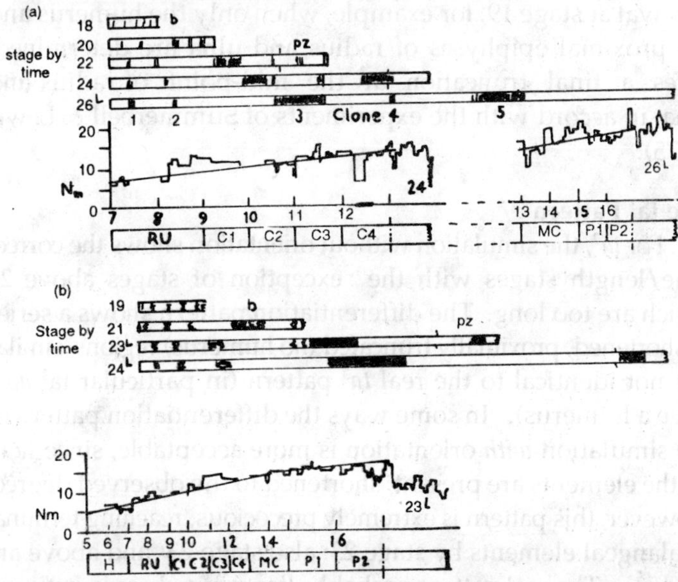

Fig. 15.6. Progress zone model for *ta*³ mutant. *(a)* Simulation without cell orientation; *(b)* simulation with cell orientation. Symbols as in Figure 15.5. Superscript 'L' indicates that the simulation is *too long* for that developmental stage (defined by time). Note the shortened proximo-distal elements that result from a wider progress zone and hence a steeper N_m profile.

The Normal Chick Pattern

The patterns in Figure 15.5 show that, as well as growing too slowly (only 150 cells long at stage 28), the 'normal' simulation without cell orientation is differentiating too slowly for the equivalent stages. At stage 28, phalangeal elements should be determined, but Figure 15.5 (a) shows only carpals. With cell orientation, however, not only is the staging by time in precise register with staging by length, but the correct stages of cell determination are also reached. This can be judged not only from the simple growth simulation but in a series of AER removal simulations. Upon removal of the AER the PZ decays over about 11 h (equivalent), giving about 0.5 N_m more determination, equivalent to a quarter of a long bone. AER removal at stage 19, for example, when only the humerus and the proximal epiphyses of radius and ulna are determined, gives a final truncation at the mid-point of radius and ulna, in accord with the experiments of Summerbell & Lewis (1975).

The ta^3 Pattern

For *ta*^{3}, the simulation without orientation shows the correct time/length stages with the exception of stages above 26 which are too long. The differentiation pattern shows a series of shortened, proximally truncated (no humerus) regions similar but not identical to the real *ta*^{3} pattern (in particular ta^3 *does* have a humerus). In some ways the differentiation pattern of the simulation *with* orientation is more acceptable, since here all the elements are present, shortened to the observed degree. However, this pattern is extremely precocious, reaching terminal phalangeal elements by stage 23: also stage 23 and above are too long. The real pattern probably lies somewhere in between these two.

Graft Simulations

Space precludes a full discussion of numerous graft simulations that have been attempted: suffice it to say that the graft of a 'young (stage 17) limb tip (10 cells) to a 'medium' (stage 21) stump (50 cells) gives *duplication*, the graft of an 'old' (stage 26) limb tip (10 cells) to a stage 21 stump gives *deletion* and the graft of an AER to a stage 21 (50 cells) stump gives a limb outgrowth with a large deletion of elements, all in accord with actual experiments. In most cases, an averaging of N_m could occur over at most 10 cells of the graft/host boundary, which is less than half a cartilage rudiment.

RELATIONSHIP TO EXISTING THEORIES

Outgrowth Induction and the Progress Zone

The model dynamics can be equated to previous theories in many ways. The orientation effect could, in both mesoderm and ectoderm, give rise to the biomechanical effects observed by Amprino and others, but as a result of gradient-controlled post-division orientation movements rather than forced moulding. The role of the AER as a sink, maintaining a region of disto-proximally increasing morphogen, and thus increasing division rate, covers both the outgrowth inductor role (Saunders) and the progress zone since it defines the point of metabolic switching. The graft of an AER in any form can, by contact, first re-initiate morphogen synthesis in competent cells, then act as a sink to establish a new region of high division rate, resulting in an outgrowth. 'Competent' cells in this context are determined but not differentiated; upon exit from the new progress zone these cells are redetermined. Present simulations indicate there should always be deletions between host level and proximal induced level.

Maintenance Factor and Senility

The model also explains the 'maintenance factor' experiments of Zwilling in that it is reasonable to suppose that the AER relies for its long-term existence on a supply of the morphogen from the mesoderm. If this is absent, as in an AER graft to incompetent flank mesoderm, the AER regresses. AP patterning could easily modify the availability of the morphogen laterally and produce an asymmetric AER: Such a concept also explains the natural regression of the AER at a stage when it is still competent to promote outgrowth. If there is a maximum attainable N_m — say 18 — after which cells *must* differentiate, then after stage 29 the supply of morphogen will decrease as cell 'drop out' of the progress zone, eventually starving the AER to death. The same experiments of Rubin & Saunders also indicate that the AER itself has only a limited life, possible based on cell division number, after which it fails as a sink and the PZ is eliminated.

During evolution these interrelations and limits have been progressively modified, together with changes in relative growth after determination to give the present range of limb patterns. It is possible that only *one* factor has changed: the rate of cell-to-cell diffusion.

Predictions of the Model

The model is not yet sufficiently refined to give specific predictions; in particular a smoother specification of N_m pattern is required, possibly based on averaging within the PZ. Within current limits, however, the correlations between simulation and reality are sufficiently close to suggest that an experimentally detectable difference in the degree of cell communication between distal strip and proximal regions should exist. There is evidence that cells differentiating to cartilage first lose much of their surface contact area but, as yet, no electrophysiological data. Similarly there should be a detectable difference between ta^3 and normal both in degree of communication and in cell orientation in the distal 500-750 µm. The latter we are currently investigating using chimaeric limb bud grafts containing small inclusions of test cells. The exact pattern to be expected in three dimensions depends on the degree of diffusion through the ectoderm and requires further investigation. Preliminary simulations indicate orientation at right-angles to the ectoderm in the surface 10-20 cells, and orientation along the PD axis in the core cells.

The pattern of determination in the model seems reasonable in the face of AER removal experiments, but predicts an even greater difference between carpal and long bone development than previously suggested. Clearly the onset of differentiation and differential growth will modify the overall pattern, but this can easily be accommodated in a 1000+ cell model that relates N_m values to modified division and orientation constants.

16

Differentiation of Animal Cell

All multicellular plants and animals formed from the union of male and female gametes begin life as a single-celled zygote. Cell division then produces a population of cells that make up the organism. There is an enormous increase in organismal complexity from egg to embryo, and this complexity allows specialized functions. Different parts of this population of cells carry out different functions: The muscles contract, the taste buds react to the chemical composition of food, the kidney cells filter wastes from the blood. The diversity of cell forms in a mammal is impressive: A conservative estimate is at least 150 cell types, each with its own distinct cellular anatomy. The definition of a distinct cell type can be somewhat arbitrary, however, because similar cell types exist in different tissues; for example, there are muscle cells that perform similar roles in your forearm and upper arm; should such cells, with a similar function but in different tissues, be considered distinct cell types? In both plants and animals, cells are organized into functional units called *tissues,* and groups of tissues in turn form *organs.* In humans there are approximately 25 organs, each made up of several to dozens of tissue types.

Within an organ tissues and their component cells are precisely arranged; this arrangement allows the organ to carry out its function. For example, the surface of the eye is a protective cornea beneath which an iris modulates the amount of light entering the eye and a lens focuses the light, the light traverses a fluid-filled cavity and then impinges on the retina, where light stimulates the sensory neurons leading down the optic nerve to the brain. In these various parts of the eye, the cells are specialized for their respective functions and hence have different shapes and in some cases different organelles. Just as important as the differentiation of the individual cell types is the complex, overall arrangement of the cells to form a functional eye. For example, in order for images to be clearly resolved when you see, the lens must be transparent to allow light to enter the eye, have the correct curvature, and be located an appropriate distance from the retina to focus light on the sensory cells.

These observations raise some of the central problems of developmental biology. How is the diversity of cell types created, how do these cells arise or migrate to specific positions at specific times during the development of the embryo, and how are the final size and shape of tissues and organ achieved? The complex problem of the regulation of cell diversification can be studied at four different levels: the molecule, the cell, the tissue, and the organism as a whole. Moreover, once a specific cell type exists, to determine how its final form tailors its function again requires the examination of cell properties at many levels. In this chapter we shall trace the diversification of several cell types, particularly blood cells, muscle cells, and a secretory cell type in the thyroid gland; as tissues are created during plant and animal development, we shall also consider how a cell's structure specifies its function.

Differentiation

Development is a gradual process, and new structures and particular cell types appear in a definite time sequence. The animal cell with the greatest potential is the fertilized egg, which has the capacity to form all the cell types of the body. This ability is termed *totipotency*. As cells in the embryo divide and interact with each other, the potential of each cell is progressively narrowed and cells become restricted to one pathway of differentiation.

In the first stage of differentiation, known as *commitment,* a cell's specific *developmental fate* is gradually set by the cell's physical position. Commitment is more of a concept than something that can be readily demonstrated by an experiment, because it is really a statement about the fate of cells if they are unperturbed. In contrast, *determination* is a series of events in the developmental history of *a group of committed cells.* For each cell within a committed group, determination involves the progressive fixation of a specific fate; when determination is complete, the cell fate is self-perpetuating and no longer requires specific interactions with the environment.

The self-perpetuating feature-of determination can readily be demonstrated by *transplantation* experiments. An example comes from transplantation of the limb buds of a chick embryo. If the area of mesoderm fated to form a fore-or hindlimb is grafted to an abnormal site on the chick embryo (such as to the somites) before limb bud outgrowth has begun, the transplanted tissue will develop into a limb. If the forelimb bud is grafted, a wing will form, but if a hindlimb bud is transplanted a leg develops. Thus the limb buds are said to be determined to be either wing or leg before they appear as recognizable structures in the embryo. Determination of tissues within each limb bud does not occur until outgrowth begins, however. If a piece of tissue from the thigh region of the early outgrowth leg bud is transplanted to the tip of the growing wing bud the transplanted tissue forms toes. That is, the tissue is determined to form leg structures (toes), as we demonstrated previously, but is not yet determined to form part of the thigh; instead it changes its fate to that of the limb region into which it is placed. Following the same procedure at a later stage of development, the thigh tissue can be seen to have become determined because the graft developed a splendid set of thigh feathers at the end of the wing. *Determination is* thus a progression of fate. Once it is fixed, the cell will continue differentiation to its final form independent of its position.

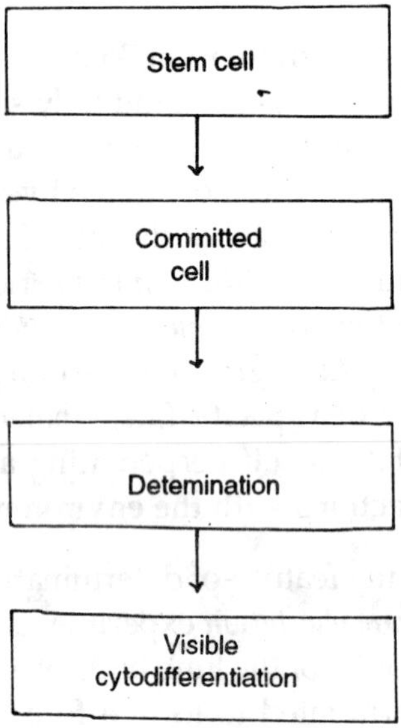

Fig. 16.1. Diagram of the stages of cell differentiation.

Although determination in animals is usually not reversible, under appropriate culture conditions a single plant cell can reform a whole plant. Such a cell is therefore totipotent and equivalent in its developmental potential to a zygote. Not all plant cells have this ability; for instance, cells of the vascular tissue cannot divide in culture, nor can they regenerate when the plant is wounded. As with animals, the developmental fate of a plant cell is also established by its position within the plant, particularly because plant cells do not move. However, many cell types of the embryo and adult can redifferentiate and assume new roles in the plant. The question of why many plant cells can reverse their differentiated state while most animal cells cannot has not yet been addressed at the molecular level.

The difference between commitment and determination is not obvious at the molecular or cellular level, although we can hypothesize that both involve some changes in gene

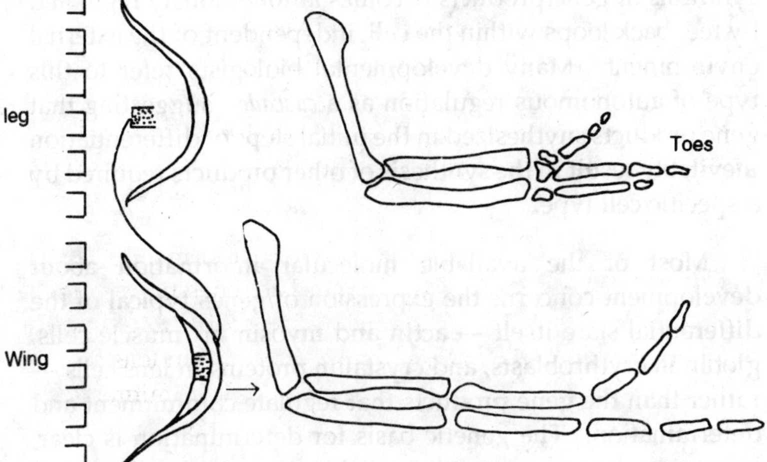

Fig. 16.2. Transplantation of thigh tissue from a chick leg bud to the tip of a developing wing bud. This region of the wing normally forms fingers, but the grafted tissue forms toes. At the time of grafting the transplanted tissue is thus determined to be leg tissue but not determined to be thigh.

expression by a cell. At the commitment stage, the expression of new genes must depend on continuous positive feedback from environmental cues and contacts with neighbouring cells. The new gene products make the cell likelier to assume a particular developmental fate; environmental cues are likely to provide additional positive reinforcement. For example, a cell can synthesize more receptors for molecules in its current environment, thus making the cell more reactive to such molecules. At present, however; our knowledge of the molecular basis of commitment is very scanty, in part because this state of the cell is so fragile. For instance, perturbing the cell's environment, including transplanting the cell, destroys the committed phenotype. Our knowledge of the molecular basis for determination is also rather weak at present , although it is clear from transplantation studies that the key feature of this state in animals is that it is self-perpetuating — that is, the synthesis of gene products becomes autonomously regulated by feedback loops within the cell, independent of the external environment. Many developmental biologists refer to this type of autonomous regulation as a *cascade,* suggesting that gene products snythesized in the initial steps of differentiation inevitably result in the synthesis of other products required by a specific cell type.

Most of the available molecular information about development concerns the expression of genes typical of the differential state itself — actin and myosin in muscle cells, globin in erythroblasts, and crystallin proteins in lens cells — rather than the gene products that regulate commitment and determination. The genetic basis for determination is clear, however, because mutants exist in which abnormal determination occurs: the best-studied cases involve the fruit fly *Drosophila melanogaster,* where in extreme instances embryos may have two dorsal sides and no ventral surface and other mutant adult flies have an extra set of wings or legs.

Stem Cells are Self-Perpetuating Cell Populations

In both plants and animals, specific populations of cells are set aside as *stem cells,* which provide a continuous supply of undifferentiated cells to replace or increase the population of differentiated cells. An essential feature of this process is that at the time of cell division, one daughter cell always remains a stem cell, while the other daughter cell is a precursor to a specific differentiated cell type, which soon starts along its specific pathway of differentiation. In effect, a *stem cell population is a self-perpetuating population of cells.* For example, hematopoietic stem cells in the bone marrow of vertebrates continuously replace the red blood cells (erythrocytes), white blood cells (granulocytes), scavenger macrophage cells, platelets, and immunocompetent cells (lymphocytes). Because the hematopoietic stem cells make only a limited range of cell types, they are considered pluripotent rather than totipotent.

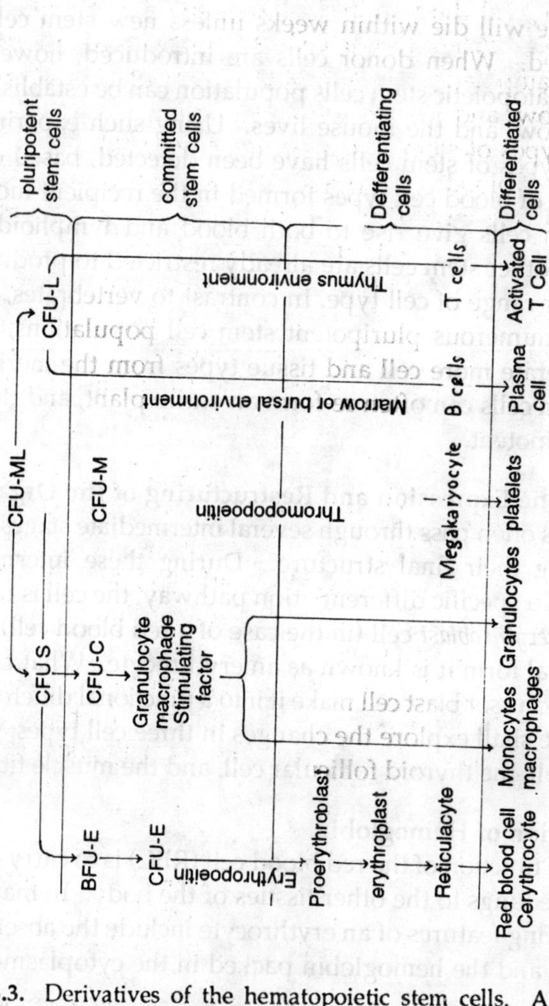

Fig. 16.3. Derivatives of the hematopoietic stem cells. A CFU is a colony forming unit. Erythropoietin, granulocyte-macrophage stimulating factor and thrombopoietin are factors involved in differentiation of cell types in a particular lineage.

A transplantation experiment can demonstrate the capabilities of the hematopoietic stem cells. First, mice are irradiated to destroy the hematopoietic cells of their marrow; because they have no means for generating new blood cells, such mice will die within weeks unless new stem cells are introduced. When donor cells are introduced, however, a new hematopoietic stem cells population can be established in the marrow, and the mouse lives. Using such experiments, several types of stem cells have been detected, based on the diversity of blood cell types formed in the recipient mouse; a few stem cells give rise to both blood and lymphoid cells, whereas other stem cells are already restricted to producing a narrower range of cell type. In contrast to vertebrates, which possess numerous pluripotent stem cell populations, plants can generate more cell and tissue types from the meristems. Meristem cells can often re-form an entire plant, and thus are truly totipotent.

New Gene Expression and Restructuring of the Organelles

Cells often pass through several intermediate stages before achieving their final structure. During these intermediate stages of a specific differentiation pathway, the cell is referred to as an *erythroblast* cell (in the case of a red blood cell), while in its final form it is known as an erythrocyte. What changes in the precursor blast cell make it into a functional differentiated cell? We shall explore the changes in three cell types: the red blood cell, the thyroid follicular cell, and the muscle fiber cell.

Production of Hemoglobin

The function of the red blood cell (RBC) is to carry oxygen from the lungs to the other tissues of the body. In mammals, the striking features of an erythrocyte include the absence of a nucleus and the hemoglobin packed in the cytoplasm.

Hemoglobin consists of a globular protein complexed to iron-containing heme groups that reversibly bind oxygen. The cell is highly specialized for its oxygen-carrying role; the RBC cytoplasm holds so much hemoglobin that it is essentially devoid of any of the organelles normally found in other cell types. A series of biochemical and structural modifications, collectively called *erythropoiesis*, transform each undifferentiated precursor cell derived from stem cell division into an immature

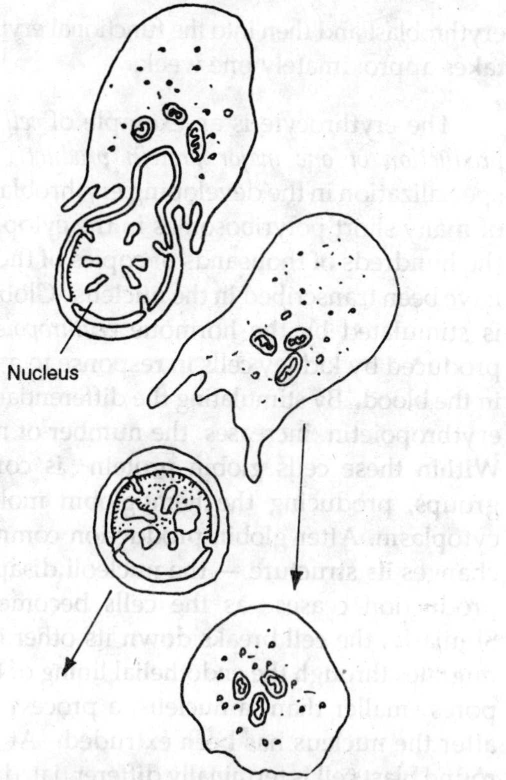

Fig. 16.4. Maturation of red blood cells involves extrusion of the nucleus. Then cells flatten into mature erythrocytes.

erythroblast and then into the functional erythrocyte; this process takes approximately one week.

The erythrocyte is an example of *cell specialization by the production of one major protein product.* The first obvious specialization in the developing erythroblast is the appearance of many short polyribosomes in the cytoplasm; they translate the hundreds of thousands of copies of the globin mRNA that have been transcribed in the nucleus. Globin mRNA synthesis is stimulated by the hormone *erythropoietin,* a glycoprotein produced by kidney cells in response to a low level of oxygen in the blood. By stimulating the differentiation of erythroblasts, erythropoietin increases the number of new red blood cells. Within these cells globin protein is combined with heme groups, producing the hemoglobin molecules that fill the cytoplasm. After globin production commences, the nucleus changes its structure — the nucleoli disappear and ribosome production ceases—as the cells become more specialized. Similarly, the cell breaks down its other organelles and then migrates through the endothelial lining of the marrow through pores smaller than a nucleus, a process that can occur only after the nucleus has been extruded. At this point the once-round blast cell is terminally differentiated to a small, flattened, disc-shaped red blood cell.

Thyroid Follicle Cells

The thyroid is a vertebrate *endocrine gland* located in humans below the Adam's apple; as with all such glands, it secretes cell products (hormones) into the blood-stream. The adult thyroid produces the hormones *thyroxine* and *tri-iodothyronine,* which regulate the rate at which cells metabolize their carbohydrate fuel. The gland originates in the endoderm of the embryo and forms as a thin pouch from the floor of the embryonic mouth. This pouch, called the *thyroglossal duct,* grows into the underlying tissues toward the neck. Its leading, closed end proliferates to produce many tubular structures; the tube between the growing gland and the floor of the mouth then disappears, thus separating the gland from the oral cavity.

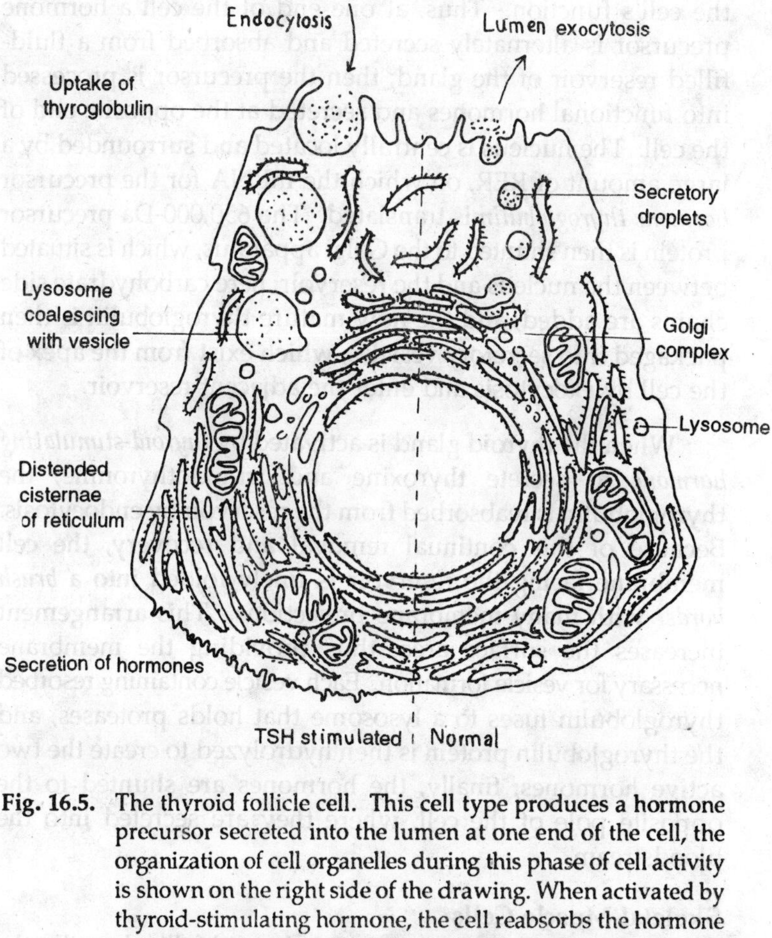

Fig. 16.5. The thyroid follicle cell. This cell type produces a hormone precursor secreted into the lumen at one end of the cell, the organization of cell organelles during this phase of cell activity is shown on the right side of the drawing. When activated by thyroid-stimulating hormone, the cell reabsorbs the hormone precursor, processes it, and secretes the mature product at the opposite end of the cell, the organization of cell organelles during this phase of cell activity is shown on the left.

In each lobe of the thyroid, a thin layer of *thyroid follicle cells* surrounds a central reservoir. The thyroid follicle cell is a good example of a *secretory epithelial cell:* Specialization of organelle arrangement within this cells cytoplasm ensures the correct production, storage, and release of the cell's hormone products. Each thyroid follicular cell has polarized cell contents, and the polarized distribution of organelles in the cell reflects the cell's function. Thus, at one end of the cell a hormone precursor is alternately secreted and absorbed from a fluid-filled reservoir of the gland; then the precursor is processed into functional hormones and secreted at the opposite end of the cell. The nucleus is centrally located and surrounded by a large amount of RER, on which the mRNA for the precursor *hormone thyroglobulin* is translated. The 650,000-Da precursor protein is then shunted to the Golgi apparatus, which is situated between the nucleus and the reservoir; here carbohydrate side chains are added, and the now-mature thyroglobulin is then packaged into secretory vesicles, which exist from the apex of the cell by exocytosis and enter the adjacent reservoir.

When the thyroid gland is activated by *thyroid-stimulating hormone* to secrete thyroxine and tri-iodothyronine, the thyroglobulin is reabsorbed from the reservoir by endocytosis. Because of this continual removal and recovery, the cell membrane facing the reservoir is finely folded into a *brush border* with many membrane projections. This arrangement increases the surface area, thus providing the membrane necessary for vesicle formation. Each vesicle containing resorbed thyroglobulin fuses to a lysosome that holds proteases, and the thyroglobulin protein is then hydrolyzed to create the two active hormones; finally, the hormones are shunted to the opposite pole of the cell, where they are secreted into the bloodstream.

Skeletal Muscle Cells

In contrast to erythrocytes and thyroid follicular cells, the *muscle cell is an example of cell differentiation in which cell fusion and the elaboration of special organelles are important features.* The precursor to a muscle fiber, called a *myotube,* is formed by the fusion of numerous mesodermal cell known as *myoblasts.* The myoblasts pool their cytoplasm and nuclei to make long, thin cells, which line up in parallel to form a developing muscle. After these cells have fused, the multi-nucleate myotubes synthesize large amounts of the two major contractile proteins, action and myosin, and several minor contractile proteins as well.

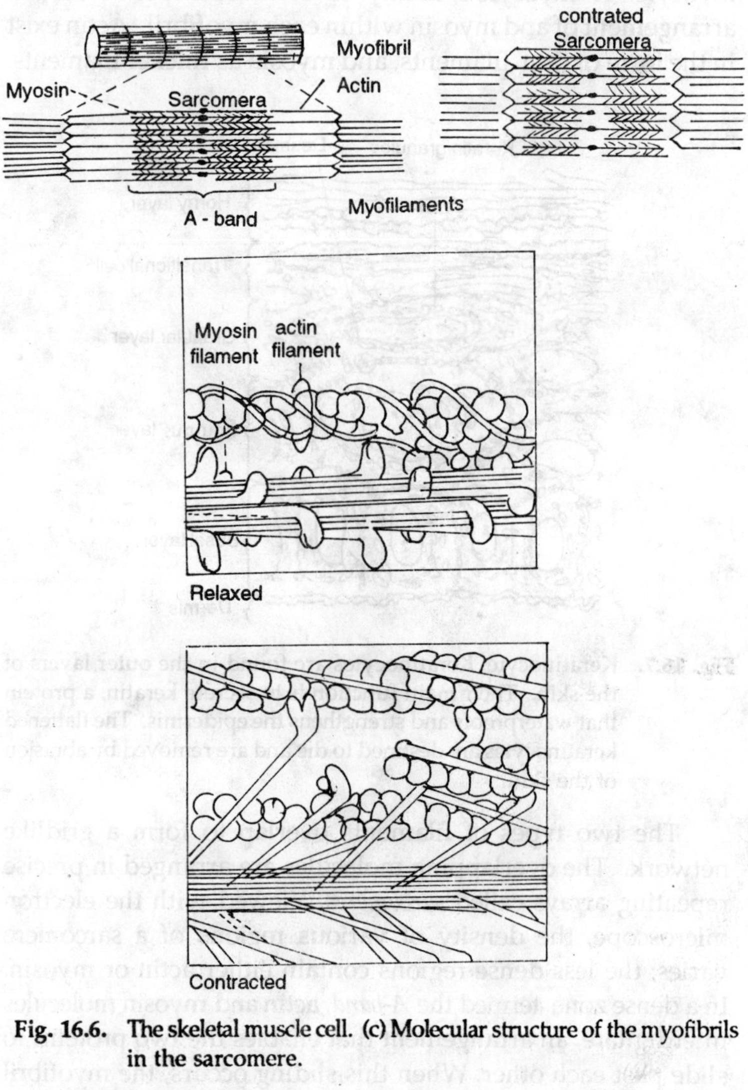

Fig. 16.6. **The skeletal muscle cell. (c) Molecular structure of the myofibrils in the sarcomere.**

An organelle found uniquely in muscle cells is the myofibril, a long, thin rod composed of the contractile proteins. Each myofibril is approximately 1 to 2 µm across and elongates by adding protein sequentially along the whole length of the myocyte. A myocyte's ability to contract depends on the arrangement of and myosin within each myofibril. Actin exist in the form of thin filaments, and myosin as thicker filaments.

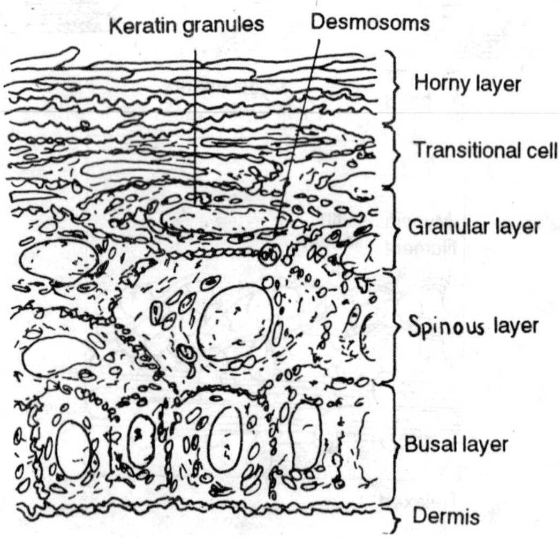

Fig. 16.7. Keratinocyte, Keratinocytes are found in the outer layers of the skin. Their main function is to secrete keratin, a protein that waterproofs and strengthens the epidermis. The flattened keratinocytes are destined to die and are removed by abrasion of the skin.

The two types of filaments overlap to form a gridlike network. The overlapping molecules are arranged in precise repeating arrays called *sarcomeres.* Viewed with the electron microscope, the density of various regions of a sarcomere varies; the less dense regions contain either actin or myosin. In a dense zone termed the *A-band,* actin and myosin molecules interdigitate, an arrangement that enables the two proteins to slide past each other. When this sliding occurs, the myofibril shortens. The sliding is co-ordinated, and because all the myofibrils contract together, the entire muscle shortens. Such

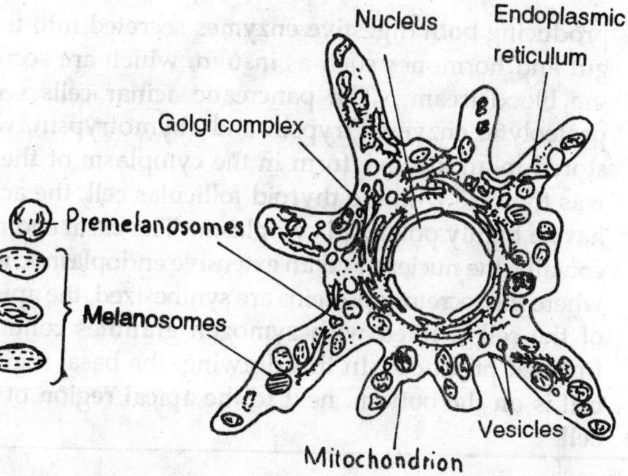

Fig. 16.8. The melanocyte. The pigment cell melanocytes are ectodermal derivatives from the neural crest. They are often elaborated branched and their cytoplasm contains melanin, a brown-black pigment that is found in a specific cell particle, the melanosome. If the melanocytes are genetically unable to synthesize melanin, the animal will be an albino. During suntanning melanosome number increases to absorb harmful ultraviolet radiation.

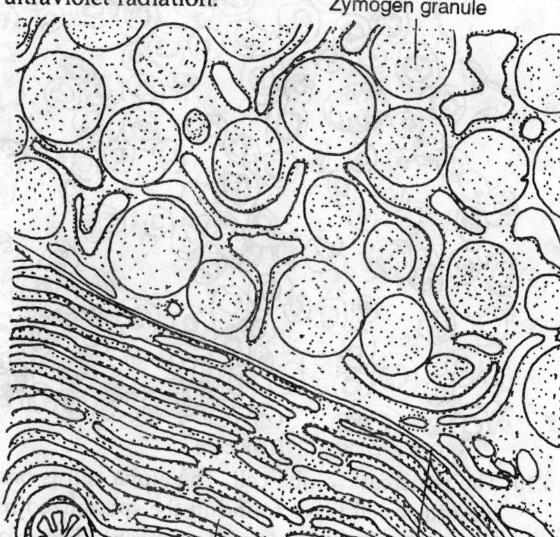

Fig. 16.9. The pancreatic acinar cell. The pancreas is a secretory organ

producing both digestive enzymes secreted into the nearby gut and hormones such as insulin, which are secreted into the bloodstream. The pancreatic acinar cells secrete the proteolytic enzymes trypsin and chymotrypsin, which are stored in an inactive form in the cytoplasm of the cell. As was the case with the thyroid follicular cell, the acinar cells have a highly polarized cytoplasm. The basal end of the cell contains the nucleus and an extensive endoplasmic reticulum, where the secretory proteins are synthesized, the apical region of the cell is filled with zymogen granules containing the finished products. In this drawing, the basal region of one cell is on the bottom, next to the apical region of a second cell.

co-ordination is achieved by the secretion of ions from a specialized ER called *sarcoplasmic reticulum,* which is wrapped around all of the sarcomeres.

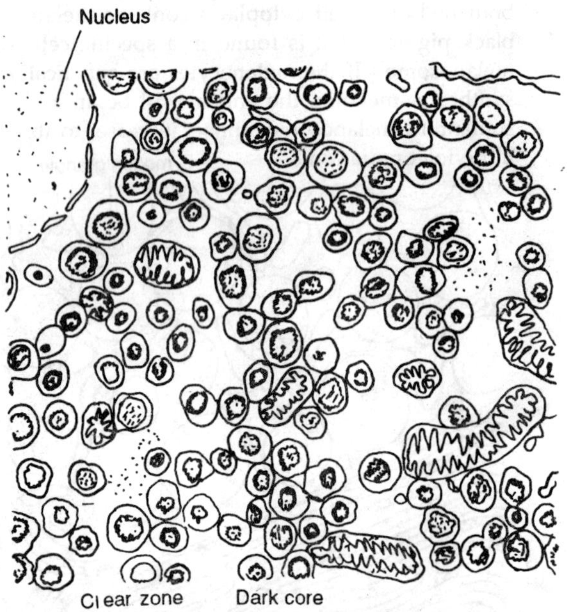

Fig. 16.10 The pancreatic alpha cell. The alpha cell is one of the four glandular cell types found in the specialized regions of the pancreas called the *Islets of Langerhans.* These four cell types each make specific hormones, which are discharged into the bloodstream through blood capillaries. The alpha cells pictured

here secrete the hormone *glucagon*. The hormone is stored in single membrane-bound vesicles, which have a dense core surrounded by a clear zone next to the membrane. These vesicles occupy much of the cell's cytoplasm, although mitochondria are also present in large numbers. In contrast to the acinar cell type, these cells have little endoplasmic reticulum.

The Arrangement of Cells Into Tissues

Cells are the basic building blocks from which tissues are made. Like the cells that compose them, tissues have specific characteristics that are suited to their functions. The point will be emphasized in this section as we examine the structure and ontogeny of three major tissues of the body: epithelia, skeletal muscle, and loose connective tissue.

The Epithelia Cover Surfaces and Form Glands

The epithelia are continuous layers or sheets of cells that cover the surfaces of the body. Most epithelia are derived from the ectoderm or endoderm germ layers of the embryo, but the mesoderm also forms some epithelia as well. The ectoderm gives rise to the epithelial portion of the skin, the largest organ system in the body. The endoderm forms the lining of the gut and its associated epithelium-lined outgrowths, such as the liver and the pancreas. A few epithelia, such as kidney tubules and the lining of the ureter, derive from the mesodermal germ layer. The special term *endothelium* is used to denote the lining of the blood vessels and some sheets of cells in the eye.

The various epithelia are specialized for a number of functions. For example, the cells lining the gut both absorb useful substances from food and secrete digestive enzymes into the gut lumen. The epidermis of the skin is designed for protection, and together with the underlying dermal skin component, the epidermis forms diverse specialized structures such as hair, nails, feathers and scales. The epithelia can also be involved in *sensory perception*: For example, sensory neurons responsible for taste are situated among the cells of the tongue epithelium.

Many of the glands formed in the embryo arise from epithelia that undergo complex changes in tissue shape and cell function. Endocrine glands, which secrete substances into the bloodstream, often lose contact with the epithelium from which they derive, as was the case with the thyroid gland discussed earlier. Likewise, the pituitary and pancreatic islet cells dissociate from the epithelium from which they arose. *Exocrine* gland secretions are transported outside the body through epithelial ducts. Familiar examples of such exocrine glands are the sweat glands of the skin and the salivary glands in the mouth.

Classification of Epithelia

Epithelia are sheets of cells often held together by desmosomes and made water-proof by tight junctions. Epithelia play a critical role from the earliest stages in animal development. The multicellular embryo is a sphere of cells, an epithelium, surrounding a cavity. Co-ordinated folding and movement of cells within this sheet shape the developing embryo. The simplest epithelial structure, only one cell thick, is appropriately called a *simple of epithelium.* Conversely, if the tissue is multilayered it is known as a *stratified epithelium;* there are also intermediate cell arrangements such as the *pseudostratified epithelium.* The shape of the cells composing epithelia also varies thus cells may be either (1) flattened, or squamous. (2) approximately square, or *cuboidal,* or (3) tall and thin, or *columnar.* Epithelia are separated from cells below by a basal lamina. This lamina is a thin, collagenous matrix to which all cells are attached in simple or pseudostratified epithelia; in stratified epithelia only the basal layer of cells is in contact with the lamina. Beneath the lamina is a thickened layer made up of collagen fibrils, which, together with the lamina, make up the *basement membrane.* The lamina alone is just 40-50 mm thick and is not visible in the light microscope.

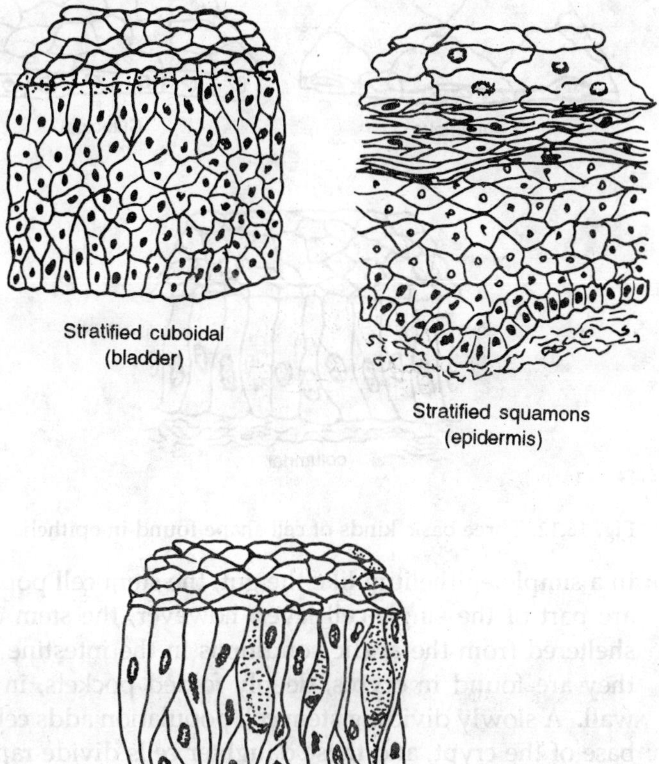

Fig. 16.11. Various types of epithelia provide the surface covering of organs.

Some epithelia are constantly abused; examples include scrapes and scratches on the skin, abrasions of the esophagus caused by the passage of large chunks of food, and the effects of digestive enzymes on the linking of the gut. Such continuous wear and tear requires *continuous replenishment* of cells. If the epithelium is stratified, the basal layer is usually a stem cell population and is the only set of cells to divide; these cells remain somewhat rounded while the daughter cells differentiate.

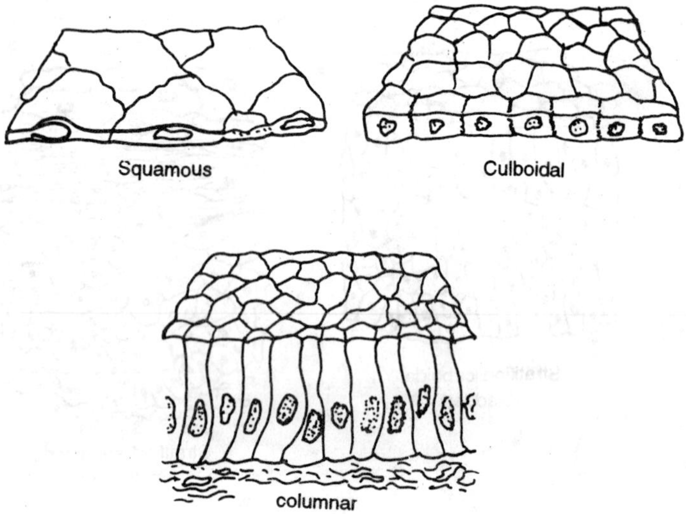

Fig. 16.12. Three basic kinds of cell shape found in epithelia.

In a simple epithelium like the gut, the stem cell populations are part of the single-cell layer; however, the stem cells are sheltered from the acidic conditions in the intestine because they are found in crypts, deeply folded pockets, in the gut wall. A slowly dividing stem cell population adds cells at the base of the crypt, and these daughter cells divide rapidly for several days, finally emerging into the lumen of the gut, where they cease division and are lost from the villus tip. In this fashion the entire lining of the intestine of a mouse is replaced approximately every two days.

The functions of epithelial cells often depend on polarized specialization on the external surface in which molecules are taken up on one side and secreted on another. Accordingly, the basal surface where epithelial cells contact the basement membrane or other cells differ from apical surfaces where epithelial cells do not contact other structures. Common apical surface specializations include cilia on tracheal (windpipe) epithelia cells, which wave rhythmically to aid the passage of particles over the epithelium. Microvilli reinforced with

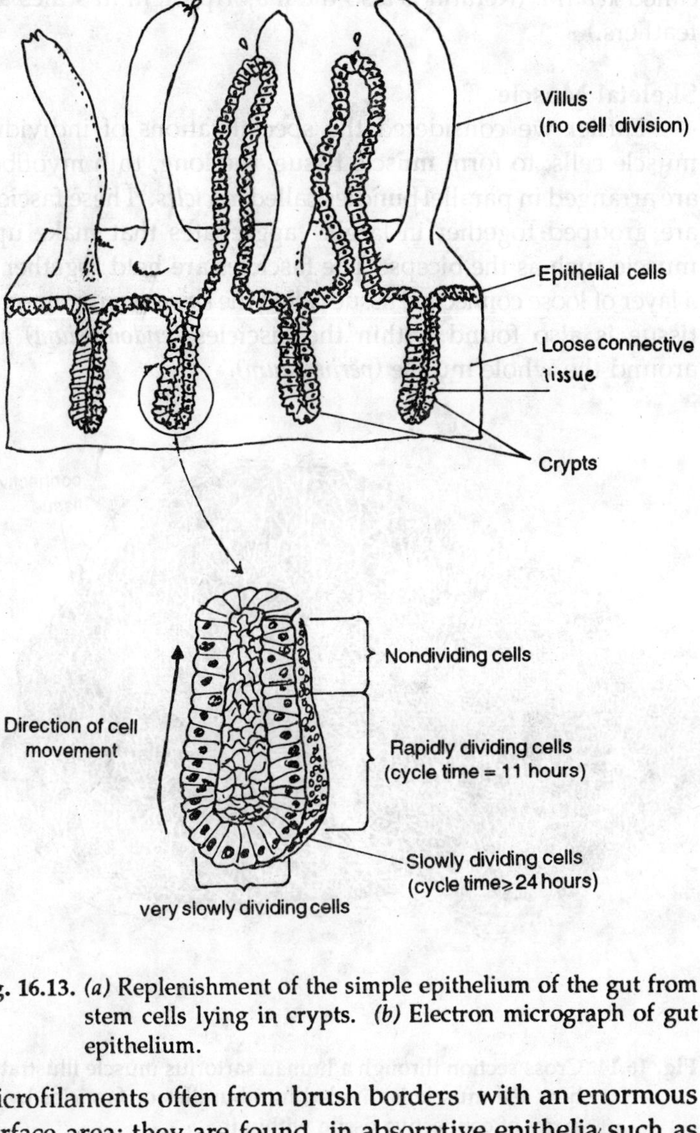

Fig. 16.13. *(a)* Replenishment of the simple epithelium of the gut from stem cells lying in crypts. *(b)* Electron micrograph of gut epithelium.

microfilaments often from brush borders with an enormous surface area; they are found in absorptive epithelia such as those of the small intestine, kidney, and the secretory surfaces of the liver. Where an epithelium serves as protection, its outer cells are often dead and filled with a fibrous protein called *keratin.* (Keratin is also the major protein in scales and feathers.)

Skeletal Muscle

Earlier we considered the specializations of individual muscle cells; to form muscle tissue, the long, thin myofibers are arranged in parallel bundles called *fascicles.* These fascicles are grouped together in larger aggregates that make up a muscle such as the biceps. The fascicles are held together by a layer of loose connective tissue called the *epimysium;* connective tissue is also found within the fascicles *(endomysium)* and around the whole muscle *(perimysium).*

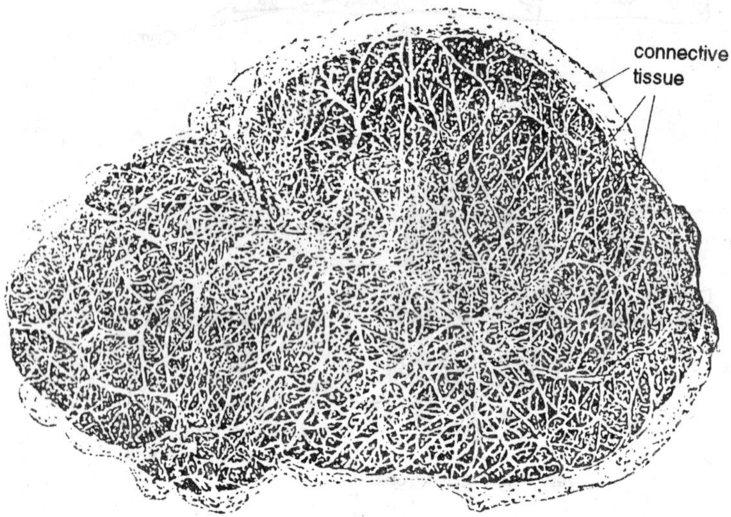

Fig. 16.14. Cross section through a human sartorius muscle illustrating how the muscle is divided into bundles or fascicles by the array of connective tissue within it.

The muscles are attached to bone by *tendons;* these very strong banks of connective tissue can stretch because of the elastic properties of the molecules composing their extracellular matrix. The longest mammalian muscle is the *sartorius*, which runs from the hip bone to the tibia just below the knee. In humans the sartorius may be up to two feet in length, while in a giraffe it can extend five feet! Muscles can also be inserted into the connective tissue layers of the skin, as is the case with facial muscles; these are the muscles that allow you to smile, frown, or wink.

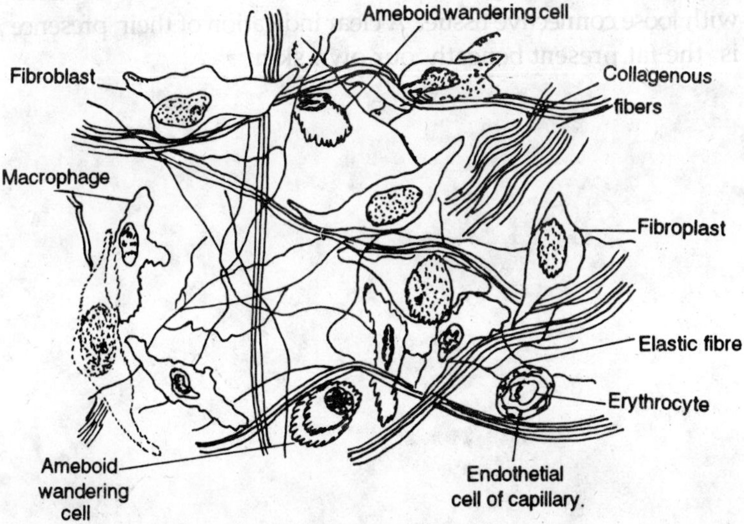

Fig. 16.15. Organization of loose connective tissue.

Connective Tissue

Connective tissue is a complex network of extracellular fibers, occasional cells, and a large population of blood capillaries. It is found throughout the body, where it supports and gives form to other tissues. Some connective tissues are dense structures composed of tightly packed arrays of extracellular matrix materials; examples include the dermis of the skin and the tendons and ligaments connecting bones and muscle groups. In contrast, loose connective tissue contains less material and is found surrounding the muscles and in *mesenteries*, the sheets of tissue in which the gut is suspended.

Fibroblasts in connective tissue produce the extracellular matrix composed primarily of collagen and elastin. Cartilage and bone are formed in areas where more rigid support is needed. The body's connective tissue also has a transient population of cells, including the scavenger cells *lymphocytes* and *macrophages,* which enter the connective tissue from the blood capillaries and remove foreign matter and dead cells. Fat storage cells known *as adipose* cells are also often associated with loose connective tissue. A clear indication of their presence is the fat present beneath our own skin.